ACKNOWLEDGMENTS

Several groups and individuals contributed to the development of the ideas in this research. First and foremost, I would like to acknowledge the seemingly unending support of Wally Milon and Clyde Kiker. Wally continued to help sculpt my high-flying ideas into manageable research even after moving on from the University of Florida. Clyde ensured that didn't loose my interest in high-flying ideas and provided much needed encouragement throughout the ordeal. The other members of my Supervisory Committee, especially Bob Emerson, are also to be commended for their timely comments and expert guidance.

Next, I would like to implicate my fellow graduate students and the group in 1094, especially Maxwell Mudhara, Bowei Xia, Mike Zylstra, Larry Perruso, Chris DeBodisco and Tom Stevens. These individuals kindly filtered many of my early thoughts on this research and provided excellent moral support. Chris DeBodisco, in particular, is to be thanked for his insights and compassion for learning.

Last, but not least, I would like to acknowledge the support of my friends and family for making the Ph.D. experience an enjoyable chapter in my life.

TABLE OF CONTENTS

LIST OF TABLES

LIST OF FIGURES

Abstract of Dissertation Presented to the Graduate School
of the University of Florida in Partial Fulfillment of the
Requirements for the Degree of Doctor of Philosophy

MEASURING THE VALUE OF PUBLIC GOODS: A NEW APPROACH WITH APPLICATIONS TO RECREATIONAL FISHING AND PUBLIC UTILITY PRICING

By

David William Carter

December 2002

Chair: J. Walter Milon
Cochair: Clyde F. Kiker
Department: Food and Resource Economics

Program evaluation (PE) techniques are adapted to measure the value of public good access. The premise is that interventions in the supply of a public good can be considered 'programs' where 'use' is tantamount to participation or 'treatment.' Three chapters (Chapters 2 through 4) explore this premise.

Chapter 2 compares the 'treatment effects' approach (TEA) to conventional revealed preference (RP) methods for valuing public good access. Program evaluation techniques are adapted to derive access value from differences in related nonmarket activity expenditures between actual and potential public good users. Unlike methods such as the travel cost approach, this approach does not estimate a structural demand or utility model to derive welfare measures. Thus, the TEA avoids many of the widely recognized problems of endogeneity in RP models. A key insight is that alternative counterfactual assumptions can be used to condition estimates of the demand for a public good.

Chapter 3 applies the TEA to measure the recreational fishing value of Gulf of Mexico petroleum platforms. Those anglers who currently fish at these platforms are the treatment group, while those who fish elsewhere are the controls. An econometric model developed in Chapter 2 is used to obtain a measure of the expected value of platform access. The measure is relatively comprehensive because the TEA readily incorporates capital expenditures. Results from the TEA model are compared to those from a travel cost model.

Chapter 4 examines the conservation value of a program that informs public utility customers about the price of service. An analytical model of perceived price is developed that can be used to assess the value of price information. The corresponding empirical models are built around a dataset of Florida water customers that assigns households to treatment and control groups based on whether or not they know the price. The results from demand and treatment effects models are used to derive the expected value of price information for an uninformed household. Importantly, the notion of counterfactuals developed in Chapter 2 allows the welfare measures to be adjusted for the possibility that price elasticities change once a household learns the price.

CHAPTER 1
OVERVIEW

The domain of consumer choice includes consumed commodities and market goods. A commodity is valued as a source of satisfaction (utility) and/or as an input into the production of a commodity that yields satisfaction. A market good is a commodity whose relative social value is given by its price in a forum of exchange (i.e., a perfectly competitive market with no externalities). These commodities are excludable and rival, that is, they are (locally) scarce and subject to competition in use. Given nonattenuated property rights for a commodity, the competition over use privileges will establish a price indicative of its relative value as a market good in a Pareto-efficient allocation (Randall 1987).

Commodities that are not market goods can be termed *nonmarket* goods. As commodities, nonmarket goods have value, but the relative value of additional units cannot be measured (directly) by an equilibrium market price. There are a number of reasons why a commodity will not be traded in a perfectly competitive market, but for present purposes the key reason relates to its *public good* characteristics. Briefly, a public good is nonexclusive and/or nonrival in consumption so that a price in use or trade cannot be established because property rights cannot be assigned. Thus, an important difference between market and public goods for consumer choice is the absence of a consistent indicator of relative value or price for the latter.

The lack of prices for public goods means that other measures must be used to evaluate the relative value of changes in the supply of these commodities. Such measures inform public policy about the potential benefits and opportunity costs of proposed changes in public good supplies (Carter, Perruso and Lee 2001). This perspective follows the long tradition of applied welfare economics, especially formal benefit-cost analysis (Johnansson 1993; Smith 1988a; Zerbe and Dively 1994). The tradition has seen the development of tools designed to recover, directly or indirectly, economic values for changes in nonmarket commodities, such as public goods. Direct inquiries require the construction of (hypothetical) nonmarket valuation transactions, whereas indirect investigations rely on the reconstruction of (actual) nonmarket valuation transactions and values based on observed market behavior (Smith 1996). This dissertation adds to the kit of so-called *revealed preference* tools that exemplify the latter indirect approach to nonmarket valuation.

Revealed Preference Valuation Of Public Goods

The research on revealed preference methods forms a vast literature documenting the attempts to recover monetary values from opportunity costs associated with observed behavior related to nonmarket and public goods. Some methods, such as the travel cost model of recreation demand, measure opportunity costs in terms of what consumers are willing to give up for access to various supplies and qualities of public goods. Other procedures, such as the averting behavior model, view opportunity costs as the amount consumers give up to compensate for a change in the supply of a public good (or bad). Still others, most notably hedonic models, consider the opportunity costs implicit in trade-offs among characteristics and prices of market goods. These approaches evolved to derive values for public goods when some of the data necessary to estimate a demand relationship is missing. Specifically, the first two approaches are attempts to impute prices for public good experiences, whereas the third approach deals squarely with a lack of data on the quantity of public goods 'purchased.'

There are a host of related problems associated with revealed preference methodologies that are common in applied demand analysis and welfare measurement. A laundry list would surely include separability, the definition of quantity and price indices, selection of functional form (with attention to choke price), recovery of compensated measures (integrability), heterogeneity and aggregation. There are a few problems such as the relationship among market, nonmarket and public goods, corner solutions, incorporation of substitutes, and the identification of income effects that have been especially troubling for revealed preference valuation of nonmarket and public good experiences. A fairly complete summary of the issues can be gleaned from Maler (1974), Johansson (1991), Freeman (1993), and Bockstael and McConnell (1999).

There is a more fundamental issue than aforementioned technical problems when attempting to value changes in a public good based on revealed preferences. A thorough welfare evaluation requires observations on consumer behavior before and after the change in the public good supply, but such panel data is rarely available. Rather, cross-section data are the norm, and evaluations of public good changes require predictions of behavior for hypothetical states of the world. The conventional approach in this case is to predict (i.e., simulate) hypothetical scenarios conditional on preference information observed either before or after the public good supply change. The difference in observed and predicted outcomes is then used to isolate welfare measures for the change in value caused by the change in the availability or configuration of the public good stock.

Alternative Approach to Revealed Preference Valuation

The approach to valuing pubic good supply changes introduced in this dissertation is based on a fundamentally different way of defining hypothetical scenarios or counterfactuals. The alternative definition arises if interventions in the supply of public goods are considered social programs. Individuals in the population who use the public good in its program (base) state are the program participants. Others who could have used the public good are nonparticipants. Then, following the literature on program evaluation (Heckman 2001b), participants are the treatment group and nonparticipants are the control group. In the tradition of the laboratory science, the net effect of the program is given by the difference in outcomes or *treatment effects* between the treatment and control groups, controlling for any inherent differences between the two groups and any (observable or unobservable) factors that may influence the participation decision. The main contribution of this research is a formal consideration of the cases in which such treatment effects can be considered measures of the value of changes in the supply of public goods. This objective is explored in three chapters. Chapter 2 develops a fairly general model that enables the use of treatment effects as welfare measures and compares this model with structural demand and utility equation approaches. Chapters 3 and 4 are applications of the principals introduced in Chapter 2.

Potential Applications

The alternative approach to revealed preference valuation can be potentially applied to evaluate public good use values in any case where the change in a public good supply can be characterized as a social program. The only crucial requirement is that observations on the behavior of participants (treatment) and nonparticipants (controls) can be clearly identified in the population of possible users of the public good. This is relatively straightforward using data typically available on explicit interventions like conservation programs at public utilities (Frondel and Schmidt 2001). The challenge lies in the identification of relevant population segments to represent participants and nonparticipants in the implicit or unintended 'programs' of agencies, Mother Nature, or human error that effectively change the availability or configuration of a public good.

Two case studies presented in Chapters 3 and 4 illustrate the potential range of applications. Chapter 3 uses the technique to evaluate a program of government intervention in the supply of artificial habitat available for recreational fishing. In this case, anglers who are observed using the habitat form the treatment group; and other potential users are considered the control group. Chapter 4 evaluates the value of a program that would fully inform public utility customers about the price of service. Here, those customers who admit knowing the price make up the treatment group, whereas all other customers are the controls. Note that the information about the price of service that is supplied by the public utility is the public good of interest in this case.

Audience

The research should be of interest to applied economists and policy makers. For applied economists, especially revealed preference researchers, the approach offers an alternative way of characterizing and analyzing the relative value of public resource allocation plans. The method offers a way to estimate the value of changes in the supply of public goods using relatively flexible demand or expenditure equations such as Engel curves. In addition, the approach can account for the possibility that preferences and/or behaviors change because of the change in the public good supply. As with any method that estimates the benefits and opportunity costs of policy proposals, the approach will add to the range of estimates available to inform policy decisions.

CHAPTER 2
TREATMENT EFFECTS AS WELFARE MEASURES

New developments are more likely when one confronts a problem with general notions of how behavioral methods work, rather than with the specific toolkit of travel cost models, defensive expenditures, etc.

—Bockstael and McConnell (1999).

Introduction

The relative value of public goods is not revealed in a competitive market. Thus, the opportunity cost of changes in public good supplies must be inferred from observations on what is, actually or hypothetically, given up to enjoy public good services. The practice of observing actual market behavior to discern the value of public goods falls under the general heading of *revealed preference methods* (Herriges and Kling 1999). These methods seek to uncover the relative value of changes in individuals' consumption mix that can be attributed to changes in public good supplies and/or qualities. This requires assumptions that separate the consumption set to isolate the purchased commodities that are *interdependent* with the public good(s) of interest (Bradford and Hildebrandt 1977; Loehman 1991).[1] Depending on the nature of the separability assumed, the ensuing analysis can focus on estimating *before* and *after* demand equations for the related individual purchased goods or for composite

---

[1] The ideas in this chapter are developed via partial analysis with assumptions regarding consumption set separability. Following Hanemann and LaFrance (1992), I acknowledge that the related welfare analysis generates partial measures of exact surpluses, but proceed in this manner to avoid the inherent ambiguities in deriving public (nonmarket) good values from incomplete systems without marginal valuation functions (Ebert 1998; LaFrance and Hanemann 1989).

commodities representing groups of purchased goods. The latter notion of separability introduces the additional complication of defining valid quantity and price indices for the composite commodities. Still further complications arise in the absence of expenditure information before *and* after the public good change.

Developing acceptable, utility-theoretic price and quantity indices for composite commodities related to public goods is especially difficult where such commodity groups are delineated according to household *activities*. For example, recreational demand models often seek to delineate composite commodities (e.g., trips) according to the location or type of recreational activities.[2] The underlying problems with this kind of commodity group delineation is readily seen when the quantity index is defined as a household production function (Blundell and Robin 2000). In this case, the price and quantity indices are fundamentally endogenous to the consumer problem and cannot be econometrically identified in a structural demand model without restrictions on preferences and/or the household production technology (Bockstael and McConnell 1981; Pollak and Wachter 1975). Despite the inherent difficulties in defining, measuring, and modeling valid quantity and price indices for activity-based composite commodities, the practice continues as somewhat of a necessary evil. For example, the pooled activity intensity, activity choice RUMs, and combined activity-intensity choice models of recreation demand all require price and quantity indices to estimate composite activity demand equations and/or (net) utility equations.

---

[2] The struggle with choice set definition in multiple-site recreation demand models illustrates the problems in delineating the consumption set according to activities (Haab and Hicks 1997; Kling and Thomson 1996; Parsons and Hauber 1998; Parsons and Kealy 1992; Parsons and Needelman 1992; Parsons, Plantinga and Boyle 2000).

The chapter begins with a review the structural approaches to valuing public good changes using observed expenditure data. Demand equation, utility equation, and combined structural approaches are covered. The review highlights the importance of the price variable in deriving public good welfare measures with each approach. Also emphasized is the way each structural approach deals with missing data on demand or utility outcomes with alternative states of a public good.

Next an alternative approach to measuring the value of access to this type of public good is introduced. The approach draws on the microeconometric program evaluation literature (Heckman 2001b) to generate uncompensated and compensated welfare measures for public good access changes without splitting out price and quantity indices from observed expenditures on a related nonmarket activity.[3] Estimators are discussed for panel and cross-section data, though, emphasis is on the latter since most revealed preference (e.g., recreation expenditures) datasets are of this type. A summary suggestions for future research concludes the chapter .

Structural Approaches to Public Good Valuation

The practical difficulties in measuring the value of public good access with observations on interdependent market goods are well-known (Bockstael and McConnell 1999). Therefore, after defining the welfare measures, I will provide only a brief sketch of approaches that focus on structural demand or utility equations. The demand equation approach characterizes a large class of methods, including the travel cost model, for analyzing the nonmarket values at the intensive margin of activity intensity. Methods

---

[3] Like most program evaluation techniques, this alternative approach is not necessarily *non-structural* (Blundell and Macurdy 1999). However, the approach is *less structural* than the demand and utility approaches requiring the estimation of structural price and quantity relationships.

following the utility equation approach are motivated by random utility theory and are generally suited to exploring values at the extensive margin of activity choice. Models that combine elements from the utility and demand equation approaches offer the potential advantage of exploring valuations at both the intensive and extensive margins in a unified discrete/continuous (D/C) choice framework. Such combined approaches can model corner solutions in the demands for the interdependent commodities *or* activities, depending on the level of analysis. Recent research in the recreational demand literature has used combined D/C approaches to address corner solutions at the activity demand level (Parsons, Jakus and Tomasi 1999; Phaneuf 1999; Phaneuf, Kling and Herriges 2000; Shaw and Shonkwiler 2000). The following discussion is meant to highlight the somewhat perplexing reliance on activity based price indices in these approaches and the way each operates in the absence of observations of behavior both before and after the public good change.

Welfare Measures

Consider the prototypical expressions of compensating and equivalent variations for a change in the condition of a public good from state *1* to state *0* in terms of the minimum expenditure function

(2-1) $$CV\left(p^1,u^1,b^1,b^0,s,\varepsilon\right)=e\left(p^1,u^1,b^1,s,\varepsilon\right)-e\left(p^1,u^1,b^0,s,\varepsilon\right)$$

(2-2) $$EV\left(p^1,u^1,b^1,b^0,s,\varepsilon\right)=e\left(p^0,u^0,b^1,s,\varepsilon\right)-e\left(p^0,u^0,b^0,s,\varepsilon\right)$$

where p is a vector of prices for market goods x, u is a utility indicator referenced to the current state of the world superscripted by *1*, b represents the supply (or quality) of a public good, and s is a vector of individual control characteristics. The term ε is a vector of stochastic elements representing heterogeneity so that there is an implicit vector of coefficients (not shown) on the variables in the model. Note that the presence of these unobservables in the expenditure functions implies that *CV* and *EV* are stochastic. Therefore, the most that can be recovered is information regarding the distribution of the welfare measures such as the expected value or some other point of central tendency. Following conventional terminology, *CV* represents the willingness to pay to prevent the change from b^1 to b^0 and *EV* is the willingness to accept compensation to allow the same change. The concepts developed in what follows are illustrated with the *CV* willingness to pay measure. Discussion of *EV* is only offered where the notion of willingness to accept offers additional insights.

A subset of the market goods are potentially interdependent with the supply of a public good. Market commodities demanded $x(u, p, b, s, \varepsilon)$ that are interdependent with the public good have generally, $\partial x(\cdot)/\partial b \neq 0$, or specifically $\partial x(\cdot)/\partial b \geq 0$ for the case of weak complementarity (WC).[4] The former relationship implies that the individual is *indifferent* to the condition of the public good when the market demands are at some minimum constant quantities (Bradford and Hildebrandt 1977), while WC places this constant minimum quantity at zero (Bockstael and Kling 1988; Maler 1974). Consequently, WC requires the additional assumption that the interdependent market

[4] The subset of the market goods that are not interdependent with the public good have $\partial x(p,u,b)/\partial b = 0$ such that public good changes only indirectly affect the purchases of these goods via income effects and the budget constraint, i.e., as $(\partial x/\partial e)(\partial e/\partial b)$.

goods are non-essential.[5] With either interdependency assumption, the value of the public good can be expressed as equivalent price changes for the interdependent commodities

$$(2\text{-}3)\qquad CV\left(p^1,u^1,b^1,b^0,s,\varepsilon\right)=\int_{p^1}^{\bar{\bar{p}}(b^1)} x\left(p,u^1,b^1,s,\varepsilon\right)dp-\int_{p^1}^{\bar{\bar{p}}(b^0)} x\left(p,u^1,b^0,s,\varepsilon\right)dp$$

where $\bar{\bar{p}}\left(b^j\right)$ is the "constant compensated demand" price vector for market goods related to b^j such that compensated demand is constant (or zero with WC) with respect to the public good (Loehman 1991).[6] The *CV* measure for the value of a change in a public good is illustrated in Figure 1 for the general interdependence and WC cases. This measure is simply the difference in the area behind two compensated demand curves. However, since demand relations are not observable as a function of utility, it is necessary to use ordinary demand equations, deal directly with utility equations, or employ some combination of demand and utility equations.

Structural Demand Approaches

Consider first strategies that rely strictly on the observed quantity demanded of the interdependent goods. The uncompensated surplus measure in terms of observable demand quantities is

$$(2\text{-}4)\qquad S\left(p^1,y,b^1,b^0,s,\varepsilon\right)=\int_{p^1}^{\bar{p}(b^1)} x\left(p,y,b^1,s,\varepsilon\right)dp-\int_{p^1}^{\bar{p}(b^0)} x\left(p,y,b^0,s,\varepsilon\right)dp$$

---

[5] The 'choke price' condition implied by WC is slightly different when the purchased good is viewed as an input in the production of an nonmarket commodity that is interdependent with the public good (Bockstael and McConnell 1983). In this case, the purchased good must *essential in the production* of the interdependent nonmarket commodity. This condition can hold even when the purchased good is not needed to produce the nonmarket commodities that are not interdependent with the public good.

[6] The double over-bar notion for the choke price vector follows Loehman (1991).

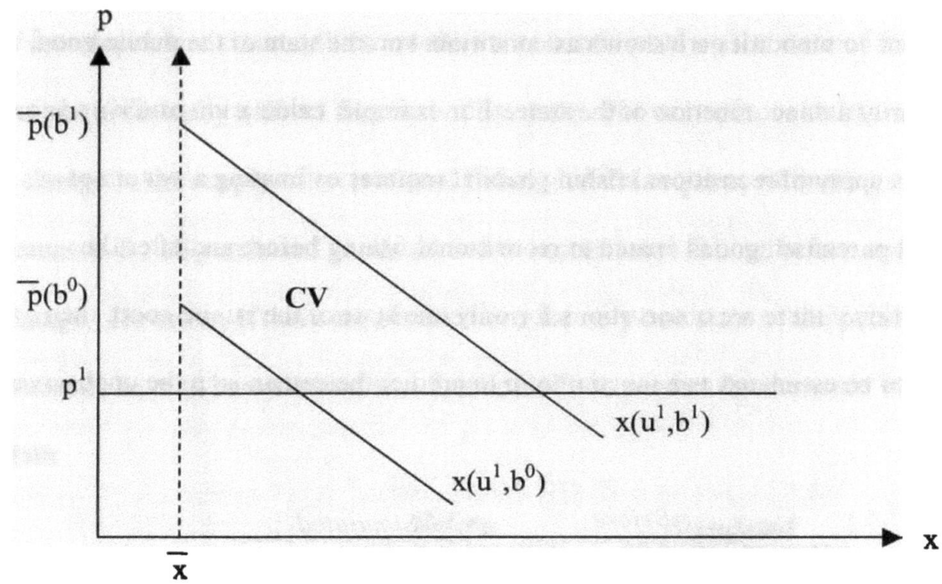

Figure 1. The value of a public good change with interdependence in demand space

where y is the constant income level and $\bar{\bar{p}}'\left(b^j\right)$ is the constant *un*compensated demand price vector. The *CV* measure equals *S* in the absence of income effects and can otherwise be recovered from (2-4) by analytically or numerically integrating back to expenditure functions with an additional (Willig) condition that rectifies the difference between $\bar{\bar{p}}\left(b^j\right)$ and $\bar{\bar{p}}'\left(b^j\right)$ (Bockstael and McConnell 1993; Hanemann 1980). The subset of these goods that are interdependent with the habitat in question define the relevant commodities to use in (2-3) and (2-4). These calculations require observations on the relevant demands before and after the change in the public good

$$\text{(2-5)} \qquad \overbrace{\begin{pmatrix} x_1\left(p,y,s,\varepsilon \middle| b^1\right) \\ \vdots \\ x_N\left(p,y,s,\varepsilon \middle| b^1\right) \end{pmatrix}}^{\textit{Before}} \quad \overbrace{\begin{pmatrix} x_1\left(p,y,s,\varepsilon \middle| b^0\right) \\ \vdots \\ x_N\left(p,y,s,\varepsilon \middle| b^0\right) \end{pmatrix}}^{\textit{After}}$$

where p references a vector of own and substitute prices and the demands for each of the N interdependent commodities are shown as conditional on the state of the public good, but not necessarily a direct function of the state. For example, calculation of CV or S for a change in the supply of recreational fishing habitat requires estimating a system of demands for all purchased goods related to recreational fishing before and after the change. However, if there are observations for only one state of the public good, then demands have to be estimated as a function of b to predict the demands in the unobserved state

$$\text{(2-6)} \qquad \overbrace{\begin{pmatrix} x_1\left(p,y,\bar{b}^1,s,\varepsilon\right) \\ \vdots \\ x_N\left(p,y,\bar{b}^1,s,\varepsilon\right) \end{pmatrix}}^{\textit{Before}} \quad \overbrace{\begin{pmatrix} \hat{x}_1\left(p,y,\bar{b}^0,s,\varepsilon\right) \\ \vdots \\ \hat{x}_N\left(p,y,\bar{b}^0,s,\varepsilon\right) \end{pmatrix}}^{\textit{Simulated}}$$

where $\bar{b}^J$ indicates the state of the public good in a manner that varies across the sample or over time for each individual. The idea is the same if data is only available after the change of interest. Once a functional form is specified for the commodity demands and the related indirect utility function, the shadow price(s) of the public good(s) can be obtained (Shapiro and Smith 1981; Shechter 1991). The value of a discrete change in the public good can be recovered by (sequentially) integrating the estimated before and after demand equations as shown in expression (2-4).

An alternative structural demand approach is to estimate a demand equation for the *interdependent activity*[7] using an acceptable quantity index to aggregate the relevant purchased quantities

---

[7] I will follow the convention of referring to the activity in which the interdependent market goods are used as the *interdependent activity*.

(2-7)
$$\overbrace{\begin{pmatrix} X_1\left(P,y,s,\varepsilon \mid b^1\right) \\ \vdots \\ X_A\left(P,y,s,\varepsilon \mid b^1\right) \end{pmatrix}}^{\textit{Before}} \quad \overbrace{\begin{pmatrix} X_1\left(P,y,s,\varepsilon \mid b^0\right) \\ \vdots \\ X_A\left(P,y,s,\varepsilon \mid b^0\right) \end{pmatrix}}^{\textit{After}}$$

where X_i is a composite commodity index for the goods used in interdependent activity $i = 1, \ldots, A$, and P is a vector of corresponding *activity-based* price indices. This is precisely what is done in the multiple site travel cost model where the 'trips' to each site (activity) are used as the quantity indices and site specific 'travel costs' are the price indices.[8] As discussed earlier, when before and after data is available the two sets of composite commodity demands can be estimated without a regressor indicating the state of the public good. In fact, with such data it is possible to take a completely nonparametric approach to recover bounds on the welfare measures using the price and quantity indices (Crooker and Kling 2000). Alternatively, two systems of composite activity demands can be estimated using the before and after data. The resulting before and after activity demand equations can be used to evaluate and recover the welfare measures (2-4) and (2-3) using price indices, instead of the prices of individual commodities. If only one set of expenditure data is available, the activity demands in the unobserved public good state have to be predicted from a demand system estimated on the observed data

(2-8)
$$\overbrace{\begin{pmatrix} X_1\left(P,y,\tilde{b}^1,s,\varepsilon\right) \\ \vdots \\ X_A\left(P,y,\tilde{b}^1,s,\varepsilon\right) \end{pmatrix}}^{\textit{Before}} \quad \overbrace{\begin{pmatrix} \hat{X}_1\left(P,y,\tilde{b}^0,s,\varepsilon\right) \\ \vdots \\ \hat{X}_A\left(P,y,\tilde{b}^0,s,\varepsilon\right) \end{pmatrix}}^{\textit{Simulated}}$$

---

[8] The single site pooled travel cost models is a straightforward simplification with only one quantity index and as many price indices as there are relevant substitutes.

As in the individual interdependent commodity demand system, $\bar{b}^j$ indicates the state of the public good. This indicator is defined in a manner that varies across the sample or over time for each individual so that the expected activity demands with changes in the public good can be simulated. If the public good indicator does not vary within each activity, then the data across activities can be combined to estimate a single demand equation. For example, when the activities are characterized as recreation sites, the data in (2-8) can be used to estimate the so-called 'pooled site' model.

Structural Utility Approaches

Now consider strategies based on random utility theory that use (conditional) utility representations. This class of models focuses on discrete events and/or activity choices involving different bundles of commodities that are interdependent with the public good. That is, the individual has (an unknown) number of ways to discretely partition (separate) their budget set to employ the services of the public good

(2-9) $$\left(y-C^{1j}\right),\left(y-C^{2j}\right),\ldots,\left(y-C^{ij}\right),\ldots,\left(y-C^{Aj}\right)$$

where $C^{ij} = p^i x^i(p, y, b^j, s, e)$ is the cost of producing alternative i using an alternative-specific subset of the purchased commodities that are interdependent with the public good. There are A such alternatives and, as before, j indicates the state of the public good. In what follows, I will use the more common convention of representing the cost in terms of activity-specific price and quantity indices $C^{ij} = P^{ij} X^{ij}$ that are conditional on the state of the public good. These price and quantity indices are analogous to the indices discussed in the structural demand approach.

Every individual implicitly has a conditional indirect utility function representing their maximum attainable utility given the activity choice and the state of the public good

(2-10) $$v^{ij} = v\left(i, y^{ij}, b^{j}, s, \varepsilon\right).$$

where the indicator i equals one if the individual chooses alternative i with the public good in state j and equals zero otherwise. The income variable is implicitly adjusted for the total spending on each alternative as $y^{ij} = y - C^{ij}$. Notice that the mix of interdependent commodities enter the problem via this virtual income term in the utility equation approach. Also note that a new dimension to the problem has been introduced. Specifically, as shown below, the utility equation approach requires additional information on activity choices to completely identify a change in an individual's utility related to a change in the public good. The four possible outcomes for each alternative are listed in Table 1.

Table 1. Utility outcomes with activity choice and change combinations

		CHOOSE ACTIVITY?	
		YES (i=1)	NO (i=0)
PUBLIC GOOD CHANGE	BEFORE (j=1)	$v\left(1, p, y^{11}, b^{1}, s, \varepsilon\right)$	$v\left(0, p, y^{01}, b^{1}, s, \varepsilon\right)$
	AFTER (j=0)	$v\left(1, p, y^{10}, b^{0}, s, \varepsilon\right)$	$v\left(0, p, y^{00}, b^{0}, s, \varepsilon\right)$

The expected unconditional utility over all alternatives is given by

(2-11) $$E\left[v\left(P, y, b^{j}, s, \varepsilon\right)\right] = E\left[\max_{i}\left\{v^{1j}, v^{2j}, \ldots, v^{ij}, \ldots, v^{Aj}\right\}\right]$$

which is presented in terms of expectations because of the stochastic, unobserved element of preferences. The value of a discrete change in the public good from b^1 to b^0 is the difference between two expected unconditional utility functions. The money metric for this value given by the value of *CV* that solves

(2-12) $$E\left[v\left(P,y+CV,b^1,s,\varepsilon\right)\right]-E\left[v\left(P,y,b^0,s,\varepsilon\right)\right]=0$$

Following Hau (1985), this value and it's money metric are depicted in Figure 2. Although not shown in the graph, note that the *i*th activity is a WC of the public good if $\partial v\left(P_{-i},\overline{\overline{P}}_i,y,b,s,\varepsilon\right)/\partial b=0$.[9] When these activities aren't chosen, utility (and expected utility) is unaffected by changes in the public good.

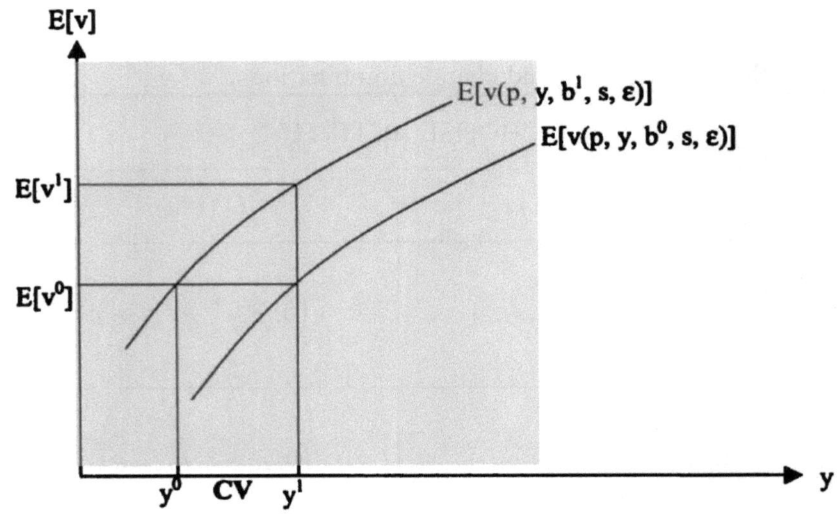

Figure 2. The value of a public good change with interdependence in utility space

Estimates of the underlying preference parameters of the before and after indirect utility equations are necessary to calculate the welfare measure in (2-12). These preference parameters have to be recovered indirectly, however, because utility is

[9] Implicitly, the *n*th purchased good is a WC of the public good if $\partial v\left(p_{-n},\overline{\overline{p}}_n,y,b,s,\varepsilon\right)/\partial b=0$. When these commodities aren't purchased, the individual is indifferent to changes in the public good.

unobservable. The standard approach involves observations on alternative choice outcomes (Hanemann 1999).[10] The present case requires data on individual choice outcomes in the before and after public good states

$$(2\text{-}13)\qquad \overbrace{\begin{pmatrix} R_1\left(P,y,s,\varepsilon \middle| b^1\right) \\ \vdots \\ R_A\left(P,y,s,\varepsilon \middle| b^1\right) \end{pmatrix}}^{Before} \quad \overbrace{\begin{pmatrix} R_1\left(P,y,s,\varepsilon \middle| b^0\right) \\ \vdots \\ R_A\left(P,y,s,\varepsilon \middle| b^0\right) \end{pmatrix}}^{After}$$

where $R_i()$ is a binary index function that equals one if alternative i is selected and zero otherwise and choices are shown as conditional on the state of the public good. The choice over multiple alternatives when the public good is in state j is motivated by a probability index model

$$(2\text{-}14)\ pr\left(R_i=1\right)=pr\left\{\left[v\left(i,P,y^{ij},b^j,s,\varepsilon\right)-v\left(k,P,y^{kj},b^j,s,\varepsilon\right)\geq 0\right] \qquad \text{for all } k\neq i\right\}.$$

where, as typically assumed, the alternatives i and k are *mutually exclusive* for the given *choice occasion*. This index can be specified once the form of the indirect utility equation is selected. Then, depending on the error structure is defined, a probability model (e.g., multinomial logit or probit) can be maximized to obtain estimates of the indirect utility equation parameters. With before and after estimates of the indirect utility function parameters, equation (2-12) can be solved to generate the value of the public good. If only one set of choice observations is available, then the alternatives have to be defined with different endowments of b

---

[10] This is also the motivation behind stated preference valuation approaches where choice outcomes are elicited for hypothetical changes in bundles of public good characteristics and individual opportunity costs (Hanemann 1984b). However, a review of stated preference methods is beyond the scope of this chapter.

$$(2\text{-}15) \qquad \overbrace{\begin{pmatrix} R_1\left(P,y,\tilde{b}^1,s,\varepsilon\right) \\ \vdots \\ R_A\left(P,y,\tilde{b}^1,s,\varepsilon\right) \end{pmatrix}}^{\textit{Before}} \quad \overbrace{\begin{pmatrix} \hat{R}_1\left(P,y,\tilde{b}^0,s,\varepsilon\right) \\ \vdots \\ \hat{R}_A\left(P,y,\tilde{b}^0,s,\varepsilon\right) \end{pmatrix}}^{\textit{Simulated}}$$

where $\tilde{b}^j$ is the public good endowment indicator. The random utility model is actually designed to handle this type of simulation based on changes in alternative attributes. A utility equation is defined with an indicator for the endowment of the public good available from each alternative activity considered. Activities can be delineated according to a public good characteristics. For example, fishing habitat can be defined by location, so that the activity of fishing in an area is uniquely (and exogenously) defined by specific public good habitat. Similarly, activities can be grouped by unique public good features that define types of an activity. For example, fishing habitat can be delineated according to whether it has man-made features so that the choice of fishing alternatives is defined accordingly (i.e., fishing artificial habitat or all other habitat). Any number of combinations is possible as well as different ways of characterizing the sequence (i.e., nesting) in which activity choices occur (Hauber and Parsons 2000; Kling and Thomson 1996; Morey, Breffle and Greene 2001). Once the alternatives have been defined, a discrete choice model can be estimated to generate utility coefficients on the public good indicator(s). These coefficients can be used to simulate the expected utility of alternative public good configurations. The welfare measure of changes the public good indicators is recovered by solving equation (2-12) using the reference and simulated expected utility levels.

Combined Structural Approaches and a Canonical Model

The unconditional demands for each activity can be defined in terms of the probability index and conditional demand functions from the two structural approaches:

(2-16) $$X_i\left(P,y,b^j,s,\varepsilon\right)=pr\left(A_i=1\right)X\left(i,P,y^{ij},b^j,s,\varepsilon\right).$$

The *X(·)* functions give the amount of each activity demanded at the *intensive* margin, conditional on the decision to *participate* and given the state of the public good. Similarly, the *pr(R$_i$=1)* functions give the probability of *participating in* an activity at the *extensive* margin given the state of the public good. When the probability of choosing an activity and the amount that is demanded are uncorrelated, then these decisions can be analyzed separately as described in previous two sections. Otherwise, these decisions should be modeled jointly in a unified corner solution model (Hanemann 1984a).[11] In this case the structural demand and utility equations come from the same consumer problem and will, therefore, share coefficient information based on shared unobservables. A structural maximum likelihood approach with cross equation restrictions is necessary to obtain unbiased estimates of the coefficient information that are shared by utility and demand equations. This approach has a long history of application to cases of nonlinear budget constraints that arise in, for example, the analysis of labor supply and the demand for public utility services (Hausman 1985; Herriges and King 1994; Hewitt and Hanemann 1995; Moffitt 1986, 1990). Structural maximum likelihood has also been used in efforts value public good changes with a combination of data on stated and

[11] There are a variety of corner solution models that have been suggested and applied to value public goods with revealed preference data (Herriges, Kling and Phaneuf 1999; Phaneuf 1999; Phaneuf, Kling and Herriges 1998).

revealed preferences (Cameron 1992). In all of these applications, the statistical problem of *self-selection* is given economic meaning and modeled accordingly.

In labor supply, the decision of how many hours to work is conditional on the decision to self-select into the workforce. When faced with a block rate pricing schedule the amount an individual demands (and their price) is conditional on the block they self-select to consume in. The number of times someone chooses to recreate at a given site or in a given activity is conditional on the decision to self-select the site or activity over all others. Note that in many of these applications, especially recreation demand modeling, self-selection is viewed as somewhat of an afterthought or a statistical nuisance. Consequently, the revealed preference methods that jointly model D/C behavior are usually concerned with *removing* the discrete outcome 'bias' from the continuous outcomes and the corresponding welfare measures. This is generally true whether the problem is addressed explicitly, as in the efforts to correct for selectivity bias in welfare measures from recreation demand models (Bockstael et al. 1990; Dobbs 1993; Laitila 1999; Shaw 1988; Smith 1988b; Ziemer et al. 1982), or implicitly, as in the recreation demand literature that seeks to derive welfare effects from unified (Phaneuf 1999; Phaneuf, Kling and Herriges 2000) or 'linked' (Parsons, Jakus and Tomasi 1999; Shaw and Shonkwiler 2000) corner solution models of participation and quantity choice. However, only the corner solution approaches give economic meaning to self-selection. In the former approach to 'correcting' selectivity bias, "the choice equation has no connection with the demand equation except for correlation in the stochastic disturbances" (Hausman 1985 p. 1262).

To conclude this discussion, consider the canonical D/C structural choice model that combines the utility and demand equation approaches

$$X\left(P,b^{j},y,s,\varepsilon\right)=\begin{cases}-\dfrac{\partial v\left(1,P,y^{1j},b^{j},s,\varepsilon\right)/\partial P_{1}}{\partial v\left(1,P,y^{1j},b^{j},s,\varepsilon\right)/\partial y} & \text{if } R_{1}=1\\ -\dfrac{\partial v\left(0,P,y^{0j},b^{j},s,\varepsilon\right)/\partial P_{0}}{\partial v\left(0,P,y^{0j},b^{j},s,\varepsilon\right)/\partial y} & \text{otherwise}\end{cases} \tag{2-17}$$

where for illustration only two mutually exclusive alternatives are considered. This model can also be written in statistical switching regime form as

$$X\left(P,b^{j},y,s,\varepsilon\right)=R_{1}X\left(1,P,y^{1j},b^{j},s,\varepsilon\right)+\left(1-R_{1}\right)X\left(0,P,y^{0j},b^{j},s,\varepsilon\right). \tag{2-18}$$

Following the discussion above, the functional forms for the utility equations in R_1 and the demand equations should embody the same representation of preferences and be estimated simultaneously with cross-equation restrictions where necessary. To recover public good values, the canonical model requires before and after data on the interdependent activities of the type in (2-7) and (2-13). With such data, utility equations or the demand equations can be used to derive the welfare measures as described in the previous sections. In the absence of before and after data, the utility or demand simulation approaches can be used to generate the values for the public good change.

Critique of Structural Approaches

As defined, the demand and utility approaches require information on the consumers' complete choice set and an indication as to those commodities that, at any point in time, are interdependent with the public good of interest. In absence of such information (or a computational method of dealing with it), the consumption set must be separated into observable/manageable components (Deaton and Muellbauer 1980).

Herein lies the difficulty with structural demand and utility approaches using the methods sketched above for measuring public good values from market data.[12]

There are at least two things to note about the use of simulation to recover welfare measures in the demand equation approach. First, using the structural demand equation(s) estimated in one state of the world to predict outcomes in another state assumes that the preference parameters will not change in response to the public good change (Whitehead, Haab and Huang 2000). That is, the simulation approach assumes that estimated structural parameters are policy invariant in the sense of the Lucas critique. In this case, the public good indicator enters as a demand shifter and the portion of the *CV* welfare measure from each interdependent commodity is simply a difference in parallel lines as shown in Figure 1. Second, the D/C choice models in the recreation demand literature allow for corner solutions in the demand for interdependent *activity-based composites*. This is different than modeling corners in the interdependent *purchased goods* that make up the composite activity-based composites. Modeling corner solutions at the activity level implicitly assumes that any interior solutions at the purchased good level before the public good change will persist after the change. However, according to Bockstael and McConnell (1993), "a discrete improvement in (a public good) can cause the individual, when maximizing utility in the new context, to choose a positive value for (an interdependent good) when previously he consumed none" (pp. 1248-1249). This means that demand system estimated on the existing data

---

[12] Hanemann and Morey (1992) show that estimates of (2-1) or (2-2) "are of no value" unless the separation is done appropriately (p. 255). I assume that separation of the consumption set is accurate in order to focus on the issues related to the price indices for groups of commodities in the separated sets. Proceeding in this manner means that all calculations of (2-1), for instance, will be a lower bound on the desired *CV* measure (Hanemann and Morey 1992).

should address the possibility of changes in the mix of purchases that occur as the public good changes. A demand system approach that models corner solutions (Lee and Pitt 1986; Wales and Woodland 1983) will be able to account for such changes at the intensive and extensive margins of purchases of the interdependent goods. This raises another issue related to the separation of the consumption set along the lines of interdependence with the public good. A complete analysis would require that the estimated demand system include *every* commodity that might be in *all* individuals' choice sets before and after the public good change. While this is clearly unrealistic, it uncovers the root cause of the intimately related problems of endogenous choice sets (see fn. 2) and endogenous prices in activity based (e.g., recreation) demand models.

The consumer *chooses* the relevant mix of purchased commodities (from the subset separated by the analyst) as part of the D/C optimization problem. In this case the marginal cost (price) of the activity will not necessarily be the same, for example, for consumers traveling from the same distance (Bockstael and McConnell 1981). This is especially problematic when attempting to simulate the value of public good changes using relationships estimated with cross-section data. With a cross-section, goods prices will not appear in goods demand equations because they do not exhibit significant variation across the sample. Thus, any variation in composite prices used in this case will be attributed to other factors. To the extent that these factors are related to preferences and not household technology (e.g., distance), price indices will misrepresent the true opportunity cost of the trip decisions.

The endogeneity problem is further exacerbated by attempts to include capital, time and the joint production of activities (Bockstael and McConnell 1981; Pollak and

Wachter 1975; Randall 1994) and coherently model corner solutions (Shaw and Shonkwiler 2000). The fact that the choice sets and activity prices are endogenously chosen by the individual make it difficult to obtain unbiased estimates of coefficients in activity level demand models. This is important because price and income coefficients are crucial in the calculation of the welfare measures. Likewise, any activity based price indices used in place of the reference and choke prices to evaluate (2-3) and (2-3) are potentially endogenous.

There are at least three ways to deal with the problem of endogenous price indices that have been explored in the literature. The first acknowledges that the price indices and derived welfare measures are ordinal measures (Randall 1994) and attempts to achieve better measures as in, for example, English and Bowker (1996). The second approach specifically models some or all of the activity prices as latent (Englin and Shonkwiler 1995) or endogenous (Fix, Loomis and Eichhorn 2000; Ward 1984) portions of the consumer problem. Models that incorporate labor supply constraints are examples of this second strategy (Larson and Shaikh 2001; Shaw and Feather 1999). A third approach attempts to choose measurement units (e.g., total distance) over which the activity can be aggregated and price indices developed in a utility theoretic manor (Shaw and Shonkwiler 2000). An alternative approach suggested in this chapter is to find ways to recover *CV* from expenditures on the related activity without the use of separate price and quantity indices.

Treatment Effects Approach to Public Good Valuation

The goal of this part of the research is to develop a welfare measure for a discrete public good change that does not require that price and quantity indices be separated from total expenditures. This welfare measure is derived from the differences in spending before and after the public good change from b^1 to b^0

$$\begin{aligned}\Delta(p^1,p^0,y,b^1,b^0,s,\varepsilon) &= p^1x(p^1,y,s,\varepsilon|b^1) - p^0x(p^0,y,s,\varepsilon|b^0) \\ &= p^1x(p^1,u^1,s,\varepsilon|b^1) - p^0x(p^0,u^0,s,\varepsilon|b^0)\end{aligned} \tag{2-19}$$

As with the structural demand and utility approaches, a separability assumption is needed to isolate the purchased commodities that are interdependent with the public good. In this case, it is reasonable to assume to Hicksian separability whereby prices in the interdependent group change by the same factor following the public good change

$$\overbrace{\begin{pmatrix} x_1(p^1_{\le L},p^1_{>L},y,s,\varepsilon|b^1) \\ \vdots \\ x_L(p^1_{\le L},p^1_{>L},y,s,\varepsilon|b^1)\end{pmatrix}}^{Before} \quad \overbrace{\begin{pmatrix} x_1(\theta_{\le L}p^1_{>L},\theta_{\le L}p^1_{>L},y,s,\varepsilon|b^0) \\ \vdots \\ x_L(\theta_{\le L}p^1_{>L},\theta_{\le L}p^1_{>L},y,s,\varepsilon|b^0)\end{pmatrix}}^{After} \tag{2-20}$$

where the first L goods are interdependent with the public good and the remaining commodities are independent. Differentiating the constant relative price expenditure function with respect to the change factor gives

$$\begin{aligned}\frac{\partial e(\theta_{\le L}p^1_{>L},\theta_{\le L}p^1_{>L},u,s,\varepsilon|b^0)}{\partial\theta_{\le L}} &= \frac{\partial e}{\partial p^1_1}\cdot\frac{\partial p^1_1}{\partial\theta_{\le L}} + \frac{\partial e}{\partial p^1_2}\cdot\frac{\partial p^1_2}{\partial\theta_{\le L}} + \cdots + \frac{\partial e}{\partial p^1_L}\cdot\frac{\partial p^1_L}{\partial\theta_{\le L}} \\ &= p^1_1x^1_1 + p^1_2x^1_2 + \cdots + p^1_Lx^1_L\end{aligned} \tag{2-21}$$

which shows that expenditure on interdependent goods can be used as a Hicksian composite commodity with the change factor as a price index (Deaton and Muellbauer 1980). A similar result holds for the $L + 1$ other commodities and the related index. Note that following the discussion of the structural models, the price index for the interdependent goods is fundamentally endogenous. In the household production context, the index represents the marginal cost of producing an activity that is interdependent with the public good.

The before and after expenditures composite for the interdependent commodities can be written in terms of the change factor price indices as

$$\overbrace{p_{\le I}^{1}x_{\le I}\left(\theta_{\le L},\theta_{>L},y,s,\varepsilon\,\middle|\,b^{1}\right)}^{before} \quad \overbrace{\theta_{\le L}p_{\le I}^{1}x_{\le I}\left(\theta_{\le L},\theta_{>L},y,s,\varepsilon\,\middle|\,b^{0}\right)}^{after} \tag{2-22}$$

and the difference in spending after the public good change can be restated as

$$\begin{aligned}&\Delta\left(p_{\le I}^{1},\theta_{\le L},\theta_{>L},y,b^{1},b^{0},s,\varepsilon\right)\\ &\quad= p_{\le I}^{1}x_{\le I}\left(\theta_{\le L},\theta_{>L},y,s,\varepsilon\,\middle|\,b^{1}\right)-\theta_{\le L}p_{\le I}^{1}x_{\le I}\left(\theta_{\le L},\theta_{>L},y,s,\varepsilon\,\middle|\,b^{0}\right)\end{aligned} \tag{2-23}$$

Similarly, the ordinary surplus and compensating variation can be written as

$$\begin{aligned}&S\left(p_{\le I}^{1},\theta_{\le L},\theta_{>L},y,b^{1},b^{0},s,\varepsilon\right)\\ &\quad= p_{\le I}^{1}x_{\le I}\left(\theta_{\le L},\theta_{>L},y,s,\varepsilon\,\middle|\,b^{1}\right)-\theta_{\le L}p_{\le I}^{1}x_{\le I}\left(\theta_{\le L},\theta_{>L},y,s,\varepsilon\,\middle|\,b^{0}\right)\end{aligned} \tag{2-24}$$

$$\begin{aligned}&CV\left(p_{\le I}^{1},\theta_{\le L},\theta_{>L},u^{1},b^{1},b^{0},s,\varepsilon\right)\\ &\quad= p_{\le I}^{1}x_{\le I}\left(\theta_{\le L},\theta_{>L},u^{1},s,\varepsilon\,\middle|\,b^{1}\right)-\theta_{\le L}p_{\le I}^{1}x_{\le I}\left(\theta_{\le L},\theta_{>L},u^{1},s,\varepsilon\,\middle|\,b^{0}\right)\end{aligned} \tag{2-25}$$

where contrary to the spending difference measure, the welfare measures hold the price level constant across states of the world. The compensating measure also holds utility constant at the level before the public good change. Consequently, if the relative prices interdependent goods remain the same after the public good change, then the difference in spending equals the surplus measure

$$\begin{aligned}\Delta^{P}&=e\left(u^{1},s,\varepsilon\,\middle|\,b^{1}\right)-e\left(u^{0}s,\varepsilon\,\middle|\,b^{0}\right)\\ &=Px_{\le I}\left(y,s,\varepsilon\,\middle|\,b^{1}\right)-Px_{\le I}\left(y,s,\varepsilon\,\middle|\,b^{0}\right)\\ &=Px_{\le I}\left(u^{1},s,\varepsilon\,\middle|\,b^{1}\right)-Px_{\le I}\left(u^{0},s,\varepsilon\,\middle|\,b^{0}\right)=S^{P}\end{aligned} \tag{2-26}$$

where $P = p^1_{\leq l} = (1)\, p^0_{\leq l}$. The price indices for the interdependent and independent goods are omitted because they equals one (i.e., $\theta_{\leq L} = \theta_{>L} = 1$) if prices are the same after the public good change.[13] A similar price constant expression can be defined for the compensating variation

$$\text{(2-27)} \qquad \begin{aligned} CV^p &= e\left(u^1, s, \varepsilon \middle| b^1\right) - e\left(u^1 s, \varepsilon \middle| b^0\right) \\ &= Px_{\leq l}\left(u^1, s, \varepsilon \middle| b^1\right) - Px_{\leq l}\left(u^1, s, \varepsilon \middle| b^0\right) \end{aligned}$$

Note also, that Loehman (1991) has shown the case of interdependent public and market goods with constant prices implies that $S^p = \Delta^p \leq CV^p$, generally, and $S^p = \Delta^p = CV^p$ if there are no income effects. These formulations are useful in when attempting to simulate the value of public good changes using relationships estimated with cross-section data. Before turning to the case of cross-section data, however, the following briefly reviews ways to recover the welfare measures from longitudinal data.

Treatment Effect Welfare Measures for Panel Data or Repeated Cross-Sections

Expression (2-26) suggests that an estimate of the uncompensated surplus measure for each individual can be recovered from panel data on expenditures before and after the change in the public good. A simple estimator of the uncompensated surplus measure expected value can be obtained by averaging $S^p = \Delta^p$ over every individual in the sample. However, this simple approach requires observations on the expenditures of each individual before and after the public good change.

In the absence of before and after expenditure data for each individual, an alternative approach can be used to recover estimates of the uncompensated measure. All

---

[13] If the prices of the goods that are not interdependent with the public good are not constant before and after the change, then the remaining economic variables can be normalized by this index to preserve homogeneity.

that is necessary is an expenditure observation for an individual when they *use* the public good and another observation when they do not. Given this information the difference in spending between use and nonuse observations measures CV^P in the special case of complete loss of public good access.

To examine this claim, note that the indirect utility levels from an activity interdependent with a public good can be divided into the four cases shown in Table 1. In this case the 'activity' is defined as 'public good use' so that a '1'denotes the utility when the public good is used and a '0' denotes the utility level otherwise. The income variable is implicitly adjusted for the total spending on each alternative as $y^{ij} = y - Px_{\leq j}\left(i, y, b^j, s, \varepsilon\right)$. Also, the price vector in the table is implicitly defined as $p = \{\theta_{\leq L}, \theta_{>L}\} = \{1, 1\}$. This sloppy notation is maintained in what follows to avoid creating another table and reduce the clutter in the functions. The corresponding four spending outcomes are listed in Table 2 where the public good use decision is explicitly labeled.

Consider the special case where b^0 represents the state of the world with no access to the public good. In this case, expression in cell (2, 2) in Table 1 and Table 2 are irrelevant because it is impossible to use the public good when access is completely restricted. The constant price difference in spending between use and nonuse of the public good with existing access level b^1 can be written explicitly as

(2-28) $$\hat{\Delta}^P = e\left(v\left(1, p, y^{11}, b^1, s, \varepsilon\right), b^1, s, \varepsilon\right) - e\left(v\left(0, p, y^{01}, b^1, s, \varepsilon\right), b^1, s, \varepsilon\right).$$

Now if we assume that an individual is just as well off without access to the public good in state b^0 as they would be if they chose not to use the public good in state b^1 then

(2-29) $$v\left(0, y^{01}, p, b^{1}, s, \varepsilon\right)=v\left(0, y^{00}, p, b^{0}, s, \varepsilon\right)$$

and expression (2-28) can equivalently be represented as

(2-30)

$$\begin{aligned}\hat{\Delta}^{p}\left(p, y, b^{1}, s, \varepsilon\right) &= e\left(p, v\left(1, p, y^{11}, b^{1}, s, \varepsilon\right), b^{1}, s, \varepsilon\right)-e\left(p, v\left(0, p, y^{00}, b^{0}, s, \varepsilon\right), b^{1}, s, \varepsilon\right) \\ &= e\left(p, v^{11}, b^{1}, s, \varepsilon\right)-e\left(p, v^{00}, b^{1}, s, \varepsilon\right) \\ &= e\left(p, u^{1}, b^{1}, s, \varepsilon\right)-e\left(p, u^{0}, b^{1}, s, \varepsilon\right)=CV^{p}\end{aligned}.$$

Using standard duality conditions (Loehman 1991) this expression can be also be written as the compensating variation in (2-1).

Table 2. Spending outcomes with public good use and change combinations

		USE PUBLIC GOOD?	
		YES (i=1)	NO (i=0)
PUBLIC GOOD CHANGE	BEFORE (j=1)	$e\left(p, v^{11}, b^{1}, s, \varepsilon\right)$	$e\left(p, v^{01}, b^{1}, s, \varepsilon\right)$
	AFTER (j=0)	$e\left(p, v^{10}, b^{0}, s, \varepsilon\right)$	$e\left(p, v^{00}, b^{0}, s, \varepsilon\right)$

To summarize, with constant relative prices and condition (2-29), the *CV* for the complete loss in public good access is given by the difference in spending on an interdependent activity for an individual when they use the public good and when they do not. A simple estimator of the compensated surplus measure expected value can be obtained by averaging $\hat{\Delta}^{p}=CV^{p}$ over every individual in the sample. However, this simple estimator susceptible to selection bias and contamination if any of the conditioning variables (i.e., *s* or *y*) change between the use and nonuse events. From assumption (2-29) an additional bias arises in this case if the conditioning variables of nonuse outcome change between the before and after states of the world. That is, if socioeconomic characteristics and/or their related parameters change when the public good changes. The parametric and nonparametric estimators reviewed in Heckman and Robb (1985) and Blundell and Costa Dias (2000) could potentially be used to correct for these biases. The application of these so-called 'difference-in-differences' and 'matching' methods to estimate the compensated measure of public good access with panel data or repeated cross-sections is a topic for future research.

Treatment Effect Welfare Measures for Cross-Section Data

With cross-section data an individual is only observed at one point in time and/or for one state of nature. That is, only one of the four possible outcomes listed in Table 1 and Table 2 is possible for any given individual in a cross-section. Consequently, each individual will have missing *counterfactual* information. For users, the counterfactual is their spending had they not used the public good. Similarly, the counterfactual for nonusers is their spending had they chosen to use the public good. From Table 1, each of these counterfactuals are possible before and after the change in the public good. To simplify matters, the approaches developed here again on the special case of constant prices and the complete loss in access to the public good from the reference access level b^1. When b^0 represents the no public good access case, there is no missing counterfactual for this state of the world because only nonuse is possible.

Conceptually, one individual's observed behavior can be used to infer about another's counterfactual and vice versa. Such inferences require information on whether or not an individual used the public good for the interdependent activity at least once for the reference period. Based on the work of Heckman and Vytlacil (2000; 2001a; 2001b), information on public good usage and two general assumptions (defined below) can be used to recover expectations of the counterfactual information missing from cross-section data.

The decision to use the public good in the reference period can be modeled with an index of net (indirect) utility

(2-31) $$D^{*}\left(p,y,b^{1},s,\varepsilon\right)=v\left(1,p,y^{11},b^{1},s,\varepsilon\right)-v\left(0,p,y^{01},b^{1},s,\varepsilon\right).$$

Following random utility discrete choice theory (Hanemann 1999), let there be an indicator variable that defines an individual's use status based on the net utility index

(2-32) $$\begin{aligned} &D^{*}=G(Z)+\varepsilon^{D} \\ &D=1 \ \text{if}\ D^{*}\geq 0, \qquad =0 \ \text{otherwise} \end{aligned}$$

where Z is a vector of all observable variables that influence the latent net indirect utility variable in (2-31) and ε^{D} is an additive error derived from ε. Note that there must be at least one variable in Z that is not in the set (s, y, p). This *exclusion restriction* is required so that we can manipulate an individual's probability of public good use without affecting their expenditures.

The two counterfactual assumptions implicit in the index model are (Heckman and Vytlacil 2001a):[14]

C1. Given that the choice probability for individuals with observed characteristics z' is $P(z')$, then if you take a *random* sample of individuals and externally set their $Z = z'$, then their choice probability is also assumed to be $P(z')$.

C2. For any case where individuals with observed $Z = z$ are set to $Z = z'$ and $P(z) < P(z')$, then: a) some individuals who would have had $D = 0$ with $Z = z$ will have

[14] Heckman and Vytlacil (2001a) also specify a series of technical assumptions that are imposed for convenience and to simplify the notation in their derivations.

$D = 1$ with $Z = z'$, and b) no individual who had $D = 1$ with $Z = z$ will have $D = 0$ with $Z = z'$

where $P(z) = Pr(D = 1 / Z = z)$ is the so-called 'propensity score' or 'choice probability' for the probability of choosing to use the public good conditional on $Z = z$. The first statement assumes that if you take a random sample of individuals and change their determinants of public good use, then the probability that they will choose to use the public good is the probability of use for those 'users' who were observed to have the same set of determinants of participation. This corresponds to assuming ε^D is independent of Z, conditional on (s, y, p), and is not essential to identify conditional expectations of the difference in spending measures (Heckman and Vytlacil 2001a). The second assumption is a monotonicity property which requires that a change to any set of factors that *increases* the probability of participation will cause some non-users to use the public good, but will never cause users to stop using the public good. The monotonicity property is implied by the additive error assumption in the index function. Both assumptions are implicit in the standard random utility discrete choice model of rational probabilistic choice (Gourieroux 2000).

I assume that the alternatives of public good use and nonuse are mutually exclusive so that the sample can be perfectly segmented into two groups based on observed behavior. To use an analogy with the program evaluation literature (Heckman 2001b), consider access to the public good in the reference state as a *program* such that public good use can be considered the program *treatment*. Those who actually use the public good make up the treatment group and all other potential users compose the *control* group. The sample will *self-select* into one group or the other.[15] The use of assumptions C.1. and C.2 allows standard selectivity methods to be used in D/C models to recover counterfactual information necessary to evaluate the welfare measures.

The approach taken in the treatment effects model departs somewhat from the conventional structural approach described above in suggesting that the information inherent in selectivity 'biases' can be used to learn about the relative value of public good access. This alternative view is not without precedent, as Heckman (2001a) notes that "evidence from self-selection decisions can be used to evaluate private preferences for the programme so that, in principle, the 'problem' of self selection can be used as a source of information about private valuations" (fn. 11).

Econometric Framework

The index function and the spending outcome equations can be jointly modeled as a D/C choice switching regression system

(2-33)
$$D^* = G\left(y,s,z \mid \beta^D,\delta,p,b^1\right)+\varepsilon^D$$
$$D = 1 \text{ if } D^* \geq 0, \qquad = 0 \text{ otherwise}$$

(2-34)
$$e^{01} \equiv px\left(p,u^{01},b^1,s,\varepsilon^{01} \mid \beta^{01}\right) \equiv px\left(y,s \mid \beta^{01},p,b^1\right)+\varepsilon^{01}$$

(2-35)
$$e^{11} \equiv px\left(p,u^{11},b^1,s,\varepsilon^{11} \mid \beta^{11}\right) \equiv px\left(y,s \mid \beta^{11},p,b^1\right)+\varepsilon^{11}$$

where β^{ij} is a conformable parameter vector for s, y and a constant such that each alternative spending outcome has its own set of parameters and an additive error term. The notation follows the earlier model where superscripts i and j denote the public good

[15] The general problem actually has two sets of self-selected samples. One set is composed of those who choose the public good at level b^1 and those who do not. The other set consists of those who choose to use the public good at level b^0 and those who do not. In the special case where $b^0 = 0$ (i.e., no public good access), there is no self-selectivity problem because there only one class of individuals: nonusers.

use decision and reference condition, respectively. Note that $j = 1$ in both outcome equations indicating that the model should be estimated with a cross-section from the period where public good access is at level b^1. Also, because I am assuming everyone in the cross-section faces the same (relative) prices and public good access level (b^1), the estimating forms of the choice and expenditure equations are conditional on these arguments. Since the indirect utility functions implicit in (2-33) and the expenditure (demand) equations in (2-34) and (2-35) come from the same consumer problem, we have $\beta^D = h(\beta^{ij}, \beta^{ij})$ and $\varepsilon^D = k(\varepsilon^{ij} - \varepsilon^{ij})$. The exact form of functions h and k will depend on the functional form selected for the expenditure (demand) equations. One form is presented in the case studies of Chapters 3 and 4. The variable z and related parameter δ are added to the index function to serve as the exclusion restriction required for the index model specification.

The spending outcome for any individual can be written as

$$\begin{aligned} e &= De^{11} + (1-D)e^{01} \\ &= px(y,s \mid \beta^{01}) + D\left[px(y,s \mid \beta^{11}) - px(y,s \mid \beta^{01})\right] + \left[\varepsilon^{01} + D(\varepsilon^{11} - \varepsilon^{01})\right] \end{aligned} \tag{2-36}$$

where the conditioning on the existing state of the public good b^1 and the constant price level is implicit. This formulation suggests that selectivity is a problem by construction if the decision to use the public good is correlated with the expenditure outcome decision. The form of the error term will differ across the observations according the specific public good use status.[16] Consequently, the data does not have the controlled (or natural) randomization necessary to identify the difference in expenditures measure as the coefficient on *D*. Selectivity correction methods aim to purge the non-random features from the data by controlling for the variation in the outcome equations due to unobservable portions of the index (choice) equation. These methods are applied in program evaluation analyses to identify moments on the distribution of *treatment effects*.

That treatment effects can be random variables is seen by rewriting expression (2-36) as

$$(2\text{-}37) \qquad e = px\left(y,s \mid \beta^{01}\right) + D\left[px\left(y,s \mid \beta^{11}\right) - px\left(y,s \mid \beta^{01}\right) + \left(\varepsilon^{11} - \varepsilon^{01}\right)\right] + \varepsilon^{01}$$

to reveal that the term multiplying the public good use indicator is a *random parameter*. Thus, the each individual can potentially have their own difference in spending treatment effect that depends on the idiosyncratic information in ε^{11} and ε^{01}. In this situation there is an underlying distribution of *heterogeneous treatment effects* and different conditioning sets will give rise to different expectations of spending differences. On the other hand, there could be only one common treatment effect parameter for all individuals given, in this case, by $px\left(y,s \mid \beta^{11}\right) - px\left(y,s \mid \beta^{01}\right)$. Heckman (1997) points to two scenarios in which treatment effects are *homogeneous* in this way. First, it can be simply be assumed that there is no unobservable portion of the expenditure differences so

---

[16] Technically, there are two ways in which the correlation between *D* and the unobservables of the outcome equations can manifest. The first way is termed *selection on the unobservables* because there is correlation between the unobservable portions of *D* and those of the outcome equations. In the other way, called *selection on the observables*, an observable element of *D* is correlated with the unobservables in the outcome equations. Note that the structural D/C modeling approaches described in the previous section generally deal with selection on the unobservables. However, a full characterization of the structural demand and utility models in terms these two types selectivity is left for future research.

that $\varepsilon^{11} - \varepsilon^{01} = 0$. Second, ε^{11} might not equal ε^{01}, but the difference between the these elements does not determine who decides to use the public good. This could happen, for example, if individuals do not know $\varepsilon^{11} - \varepsilon^{01}$ at the time they choose to use the public good and their best guess of it is zero. The expected value of the net effect of unobservables becomes zero if these individuals' expectations of $\varepsilon^{11} - \varepsilon^{01}$ are typical of the entire population. If either of these scenarios is true, then there is no distribution of spending treatment effects and there is only one expected treatment effect measure for the population. The following analysis assumes the more general heterogeneous treatment effect case so that I can tailor mean spending difference parameters that correspond to the various welfare measures discussed so far.

Treatment Effect Welfare Measures

There are two ways in which the econometric framework can be used to recover a population level measure of $E\left[CV^p\right]$. The first recovers $E\left[\hat{\Delta}^p\right] = E\left[CV^p\right]$ using assumption (2-29) and additional assumptions based on the spending behavior of those who use the public good and those who do not for the interdependent activity. The second way does not require condition (2-29) and instead seeks to recover $E\left[\widehat{\Delta^p} \middle| M\right] = E\left[CV^p\right]$ using an indifference set M to condition the distribution of differences in spending with and without public good use. This second approach aims to develop a *policy relevant* difference in spending measure (Heckman and Vytlacil 2001b).

Approach 1. With constant relative prices the uncompensated surplus S^p for the total loss of public good access is measured by the difference in spending by an individual when they use the public good use and when they do not. For the model in (2-33)-(2-35) this difference in spending for each individual is given by

(2-38)
$$\hat{\Delta}^{p} = px(y,s \mid \beta^{11}) - px(y,s \mid \beta^{01}) + (\varepsilon^{11} - \varepsilon^{01})$$

where the conditioning on b^{I} and p are again left implicit. This is the heterogeneous treatment effect random parameter defined in expression (2-37). There are three commonly used measures of the expected value of this variable using different sets of the sample (Heckman and Vytlacil 2000). The unconditional expected value measures the so-called *average treatment effect*

(2-39)
$$ATE^{\hat{\Delta}^{p}} = E\left[\hat{\Delta}^{p} \middle| y,s\right] = E\left[px(y,s \mid \beta^{11}) - px(y,s \mid \beta^{01}) \middle| y,s\right]$$

This mean measures the expected difference in spending from public good use for a randomly chosen individual. If corrected for selectivity, the *ATE* will approximate the mean treatment effect from a randomized experiment.[17] Evaluating the expected value of the treatment effect over the support of those who chose to use the public good gives the effect of the *treatment on the treated* as

(2-40)
$$TT^{\hat{\Delta}^{p}} = E\left[\hat{\Delta}^{p} \middle| y,s,D=1\right] = E\left[px(y,s \mid \beta^{11}) - px(y,s \mid \beta^{01})\right] + E\left[\varepsilon^{11} - \varepsilon^{01} \middle| D=1\right]$$

A similar parameter can be defined for the segment of the sample who chose not to use the public good

---

[17] There are two ways in which selectivity can bias the experimental treatment average (Winship and Morgan 1999). The mean selection bias given by $E[\varepsilon^{01} \mid y,s,D=1] - E[\varepsilon^{01} \mid y,s,D=0]$ indicates how spending in the reference level of the public good differs between users and nonusers. The second source of bias occurs if the change in spending caused by public good access/use (treatment) is different among users and nonusers. This bias is given by $E\left[\hat{\Delta}^{p} \middle| y,s,D=1\right] - E\left[\hat{\Delta}^{p} \middle| y,s,D=0\right] = E\left[\varepsilon^{11} - \varepsilon^{01} \middle| y,s,D=1\right]$. Neither of these spending differences can be attributed to public good access.

(2-41)

$$UT^{\hat{\Delta}^p} = E\left[\hat{\Delta}^p \middle| y,s,D=0\right] = E\left[px\left(y,s \mid \beta^{11}\right) - px\left(y,s \mid \beta^{01}\right)\right] + E\left[\varepsilon^{11} - \varepsilon^{01} \middle| D=0\right]$$

Using the counterfactual assumptions in C.1 and C.2, the above treatment effects measure the expected value of S^p for the loss of public good assess for different segments of the population. In the previous discussion, I suggested that, with panel data, condition (2-29) can be assumed so that $\hat{\Delta}^p = CV^p$ for each individual. With cross-section data, however, condition (2-29) alone is not sufficient for $\hat{\Delta}^p$ to measure CV^p because there is additional missing counterfactual information. Recall that this condition requires that an individual is just as well off without access to the public good in hypothetical state b^0 as they would be if they *chose* not to use the public good in the reference state b^1. With a complete panel there are observations on expenditures for each individual when they choose to use the public good and otherwise for the interdependent activity (with the public good fixed at b^1). With a cross-section, however, there is only information from nonusers about spending and the corresponding utility level when the public good is not used. Similarly, only users provide information about spending and the corresponding utility level when the public good is used. This means that a direct application of condition (2-29) with cross-section data will involve an interpersonal comparison of well-being. The extent of the comparison will depend on which treatment effect measure is used.

For $ATE^{\hat{\Delta}^p}$, $TT^{\hat{\Delta}^p}$, or $UT^{\hat{\Delta}^p}$ to measure the E[CV^p] for complete loss of access, condition (2-29) needs to hold for those who didn't choose to use the public good at the reference state. In addition, for $TT^{\hat{\Delta}^p}$ to measure E[CV^p] for users we require

(2-42) $$\left[v\left(0,y^{01},p,b^1,s,\varepsilon\right) \middle| D=1\right] = \left[v\left(0,y^{01},p,b^1,s,\varepsilon\right) \middle| D=0\right]$$

for those who did choose to use the public good in the reference state. This assumption requires that the indirect utility level of users (D=1), had they not used the public good, is the same as nonusers (D=0) with the same set of characteristics (*y, s, ε*). Similarly, to use $UT^{\hat{\Delta}^p}$ as a measure of E[CV^p] for nonusers we additionally require

(2-43) $$\left[v\left(1, y^{11}, p, b^1, s, \varepsilon\right)\middle|D=0\right]=\left[v\left(1, y^{11}, p, b^1, s, \varepsilon\right)\middle|D=1\right]$$

for those who did not choose to use the public good at the reference state. This assumption states that the indirect utility level of nonusers (D=0), had they used the public good, is the same as users (D=1) with the same set of characteristics (*y, s, ε*). In order for $ATE^{\hat{\Delta}^p}$ = E[CV^p] for the entire population, we need assumptions (2-42) and (2-43) as well as assumption (2-29).

Approach 2. To motivate the task of specifying and identifying the policy relevant treatment effects measure, consider the distribution of expenditure differences $\hat{\Delta}^p$ shown in Figure 3 where *f()* is a function describing the relevant density of ε.[18] The unknown switching threshold is shown as $\hat{\Delta}^{p*}$ which also corresponds to the compensation that would make the *marginal individual* just indifferent between using or not using the public good. Individuals located to the right of the threshold *choose* to use the public good alternative whereas those to the left do not. Thus, there is actually a related distribution of $\hat{\Delta}^{p*}$ for the marginal *and* non-marginal individuals. If the public good does not have value to the individual outside its use in the interdependent activity, then

[18] Following Moffitt (1998) the distribution of the treatment effect in can be depicted as in Figure 3 by assuming that the choice between alternatives is based entirely on $\hat{\Delta}^p$ and that this 'selection' is positive: individuals with high values of $\hat{\Delta}^p$ are relatively more likely to choose the first alternative than those with lower values.

$\hat{\Delta}^{p*}$ corresponds with the CV^p measure. Similarly, the expected value of $\hat{\Delta}^{p*}$ for the population corresponds with the E[CV^p] measure.

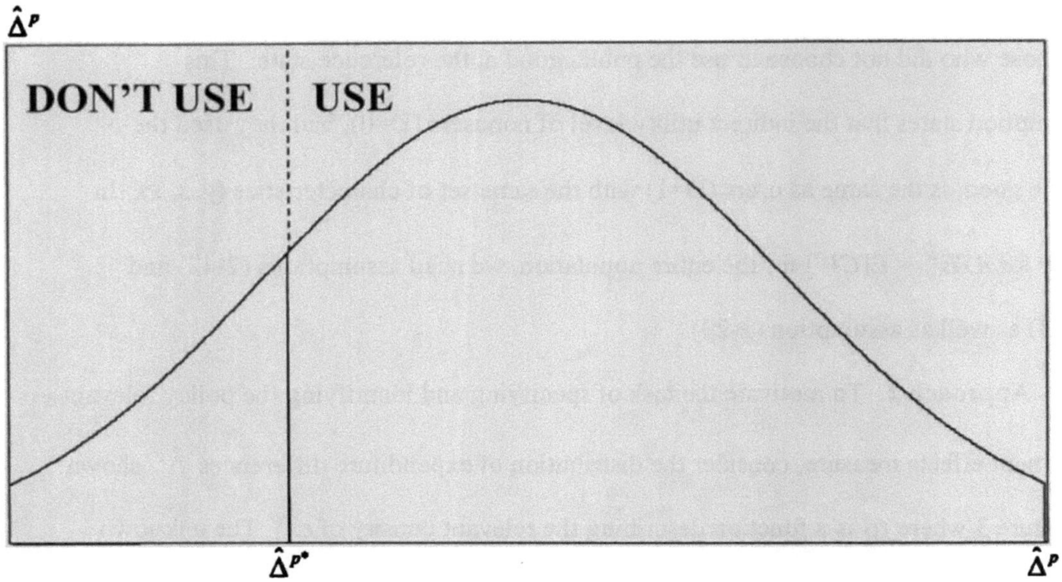

Figure 3. Expenditure difference threshold

Referring again to Figure 3, the unconditional mean of $\hat{\Delta}^{p}$ will not correspond to an exact welfare measure of public good access (unless the conditions specified in Approach 1 for $ATE^{\hat{\Delta}^p}$ are met). The mean of the treatment effect has to be conditioned in order to recover a value that represents the central tendency of the exact measure $\hat{\Delta}^{p*}$. The expected value of $\hat{\Delta}^{p*}$ can be thought of as the expected value of $\hat{\Delta}^{p}$ conditional on being at the point of indifference for each individual or

(2-44)
$$\begin{aligned} E\left[\hat{\Delta}^{p*}\right] &= E\left[\hat{\Delta}^{p} \middle| v\left(1,p,y^{11},b^{1},s,\varepsilon\right) = v\left(0,p,y^{01},b^{1},s,\varepsilon\right)\right] \\ &= E\left[\hat{\Delta}^{p} \middle| D^{*} = 0\right] \end{aligned}$$

This type of treatment effect is known as the marginal treatment effect (MTE) or the limit version of the so-called local average treatment effect.[19] According to Heckman and Vytlacil (2001a), the MTE "has the interpretation as a measure of willingness to pay on the part of people on a specified margin of participation in the program" (fn. 16). The conditioning expressions in this case can be viewed as the indifference set (Heckman 1997). Using the definition of $\hat{\Delta}^P$ in (2-38) and the additive-error index function specifications in (2-33), this expectation can be written as

$$\begin{aligned} E\left[\hat{\Delta}^{P*}\right] &= E\left[\hat{\Delta}^P\right] + E\left[\varepsilon^{11} \middle| \varepsilon^D = -G(y,s,z)\right] - E\left[\varepsilon^{01} \middle| \varepsilon^D = -G(y,s,z)\right] \\ &= E\left[\hat{\Delta}^P\right] + \left(\sigma^{11D} - \sigma^{01D}\right)\left[-G(y,s,z)\right] \end{aligned} \tag{2-45}$$

where $E\left[\hat{\Delta}^P\right]$ is the unconditional average treatment effect from (2-39) and the conditioning set for the index function $G\left(\cdot \mid \beta^D, \delta, p, b^1\right)$ is omitted to simplify the notation. The terms σ^{01D} and σ^{11D} measure the covariance between the public good use decision and the unobservable portion of each expenditure outcome.[20] These covariances show how, for a given set of set of (y, s, z), spending on the interdependent activity change with a change in the net utility of public good use. Information about an individual's preferences can be recovered by examining the signs of (σ^{11D} - σ^{01D}) and $G(\cdot)$.[21] If these terms have opposite signs, then the propensity to change spending is greater with public good use than without. In this case, an individual prefers to use the

---

[19] Attribution for this treatment effect parameter is given footnote 18 of Heckman and Vytlacil (2001a).

[20] Note that the covariances arise in the derivation from the general expression for the expectation of a random variable conditional on another random variable, i.e., $E[\varepsilon^{ij} \mid \varepsilon^D] = \sigma^{ijD}\varepsilon^D$.

[21] The economic interpretation of the switching regression covariances is similar to the use of these parameters in the labor literature that examines comparative and absolute advantage (Dolton and Makepeace 1987; Emerson 1989).

public good, otherwise they do not. Furthermore, since the first term on the right-hand side is simply a difference in expenditure, the remainder has the interpretation as the *additional* amount necessary to make the individual indifferent with and without public good use. This becomes apparent when we recognize that σ^{11D} and σ^{01D} give the slope of lines that showing how the expenditures of users and nonusers, respectively, vary with the net utility of public good use indexed by $G(\cdot)$.[22] The term $G(\cdot)$ for each individual gives the value of the unobservables necessary to maintain the same utility level (i.e., D*=0) with and without the public good (use). The constant covariance term translates this amount into the money measure of the additional compensation necessary to maintain the same utility level when the public good (use) is not available.

The relationship among the treatment effect measures of spending differences in (2-39)-(2-41) and the exact measure in (2-45) can be used to formulate a model of public good use participation. The expected value of the exact measure $E\left[\hat{\Delta}^{P}\right]$ represents the mean threshold of public good use for the sample. It is straightforward to show that participation is expected on average if

(2-46) $$E\left[\hat{\Delta}^{P*}\right] \geq ATE^{\hat{\Delta}^{P}}$$

because this implies that the utility of using the public good is greater than the utility otherwise.[23] The corresponding expressions for the users and nonuser groups are, respectively,

---

[22] The with and without public good (use) expenditure equations could be drawn on the same graph as a function of net utility. This may provide a useful way of visualizing the treatment effect welfare measure(s). I leave this for future research.

[23] From (2-26) and (2-27), $CV^{P} \geq \Delta^{P} \Rightarrow e(u^{1},p,b^{0},s,\varepsilon) \geq e(u^{0},p,b^{0},s,\varepsilon)$ which holds if $u^{1} \geq u^{0}$ (all else equal) because the expenditure function is increasing in utility. Note that u^{1} is the utility with the public good (use) and u^{0} is utility otherwise.

(2-47) $$E\left[\hat{\Delta}^{p^*} \middle| user\right] \geq TT^{\hat{\Delta}^p}$$

(2-48) $$E\left[\hat{\Delta}^{p^*} \middle| nonuser\right] \geq UT^{\hat{\Delta}^p}$$

These conditions provide a consistency test on the treatment effects model results. We would expect condition (2-47) to be true for those who actually choose to use the public good for the interdependent activity. However, the inequality in condition (2-48) is expected to be reversed because this expression applies to the group of individuals who choose not to use the public good.

Discussion

This chapter introduced the treatment effects approach to evaluating the welfare effects of changes in the condition of public goods. The approach applies techniques from the program evaluation literature to develop measures of welfare changes from spending on market goods that are interdependent with a public good. This approach views interventions in the supply of public goods as *programs* where the segment of the population currently using these goods are viewed as the *treatment* group and other potential users are considered the *control* group. Measures of the value of public good access are recovered from differences in expenditures among users and nonusers. This approach offers the advantage of using price constant specifications of demand relations (e.g., Engel curves) because the typical 'choke' price argument is not required to evaluate the access restrictions.

There are several directions for future research on the treatment effects approach. First, the model can be extended to evaluate continuous treatment effects (Heckman 1997) to deal with a continuum of possible changes in public good conditions. This would also allow for a richer consideration of the counterfactual assumptions required for cases other than with and without public good access case considered in this chapter. Second, there is more work necessary to examine the role of substitutes in the treatment effects approach. The importance of substitutes other revealed preference approaches such as the travel cost model is well-documented (Kling 1989; Rosenthal 1987; Smith 1993). Recent research on program evaluation (Heckman, Hohmann and Smith 2000) suggests methods to account for substitute programs in estimates of treatment effects that may be useful in generalizing the approach introduced in this chapter. A related direction for further research is to examine models for multiple programs that could be used as a treatment effects analog to the multi-site travel cost model.

Fourth, future work on the treatment effects approach should include applications the nonparametric estimators developed in the program evaluation literature to the public good valuation problem. Finally, the longitudinal measures introduced here should be explored further. This would require a panel or repeated cross-sections of expenditures on activities interdependent with a public good. Although the former is relatively rare, repeated cross-sections are regularly collected by variety of resource management agencies.

CHAPTER 3
APPLICATION TO RECREATIONAL FISHING

There are approximately 4,000 offshore oil and gas structures in the state and federal waters of the Gulf of Mexico. These structures account for a major proportion of the available fish habitat in the Gulf and they are utilized by a variety of recreational users (Quantech 2001). More than 100 structures are removed annually but the U.S. Mineral Management Service has adopted a "Rigs-to-Reefs" policy to mitigate the loss of these structures to maintain the public good benefits of fisheries habitat (Dauterive 2000). This policy involves leaving the structures in place, toppling the structures to create benthic habitat, or moving them to a new location. While the costs of removal are relatively well known, the economic benefits of current usage and of retaining these structures have not been estimated.

This chapter presents an analysis of the value of access to petroleum rigs for recreational fishing. It is hypothesized that fishing at offshore rig sites requires additional fishing capital compared to other types of angling. Consequently, the analysis measures opportunity costs in terms of per trip costs *and* expenditures on fishing capital. This requires that models of recreational fishing, such as the travel cost model, be adapted to jointly model choices over durable and nondurable goods.

Randall notes that in applications of the travel cost model the "allocation of the costs of owning and maintaining vehicles and other durable equipment to any particular trip (activity) proceeds, if at all, in an arbitrary fashion" (p.90). This is largely due to the additional complications that arise when attempting to introduce (joint) capital expenditures into fundamentally endogenous price indices for nonmarket activities (Bockstael and McConnell 1981; Pollak and Wachter 1975). From Chapter 2, such price indices are necessary in the travel cost model in order to estimate a demand equation for the nonmarket activity and derive welfare measures of access or quality changes. However, it may not be possible to derive valid quantity and price indices when 'trips' are used as the aggregator for recreation commodities.[1]

Linear random utility models also typically ignore capital expenditures because it is presumed that these expenditures do not vary with the number of visits to a particular site or type of site. In this case, capital stock or expenditures are individual-specific variables that drop out of the model because the estimation is based on utility differences and these variables do not vary across sites or activities. This may be a valid assumption when (perfect) substitute sites or activates are available that jointly use the durable equipment; that is, when expenditure categories are not uniquely related to the attributes of an activity or site. In other cases, however, ignoring such expenditures could seriously misstate welfare estimates for policies that stand to affect the access to, or quality of capital-intensive activities.

This chapter presents two approaches to incorporating annual capital expenditures into estimates of the value of access to petroleum rigs for recreational fishing. The first approach is a simple adaptation of the structural travel cost model to incorporate the stock of fishing capital among explanatory variables. Like the conventional travel cost model,

---

[1] Shaw and Shonkwiler (2000) demonstrate that the price indices suggested for 'trips' in the literature are not valid or that these indices do not enter the trip demand equations in a valid way. A valid price index for a Hicksian composite commodity is linear homogenous in goods prices and the specification of the trip demand equation is homogeneous of degree zero in income and the price index.

this adaptation requires that price and quantity indices be separated from fishing expenditures to estimate a structural model of fishing trip demand. The second approach is based on the treatment effects framework for measuring public good values presented in Chapter 2. This framework involves conditions whereby welfare changes can be measured by differences in the observed expenditures of two segments of the population who participate in recreational fishing: petroleum rig users and nonusers. The method can be implemented with raw expenditures on a recreational fishing activity that is interdependent with access to petroleum rigs. Price and quantity indices do not need to be separated from the aggregate annual expenditures for each individual. Data used in the analysis are drawn from intercept and phone surveys of marine recreational anglers along the Gulf of Mexico coast (Alabama to Texas) that elicited detailed information about site-specific activities and expenditures for variable and capital goods *directly* related to the activity. The econometric estimation procedure developed in Chapter 2 controls for (and actually takes advantage of) activity specific selectivity, in this case, the choice whether to fish near an petroleum structure or not.

Welfare Measurement with Capital Expenditures

There are few, if any, attempts to systematically incorporate expenditures on durables into models designed to measure the value of changes in public goods. Studies frequently use indicators of existing capital stock as explanatory variables in demand equations. Travel cost analyses of recreational fishing, for example, often incorporate dummy variables for boat ownership. However, the rationale for including such variables is usually not fully developed beyond an implicit notion that the behavioral relationships estimated are *conditional* on the existing stock of capital (Pollak 1969). Capital stock indicators are included among regressors to *control* for variations in holdings in demand

and welfare calculations. This formulation treats capital as an *exogenous* portion of the consumer problem and is correct to the extent that the additions to capital are fixed over the (decision) period of interest. In many cases, though, the stock of capital is better characterized as 'quasi-fixed' so that periodic increments are chosen along with (non-capital) commodities as part of the same optimization process (Conrad and Schroder 1991).

Consider that the choices regarding where and how frequently to fish annually are conditional on boat ownership, but that a boat could be purchased at any point during the year. In this example, the same observable and unobservable factors that influence the choice of capital stock levels also determine the demand for other commodities. Consequently, measures of *current* capital stock and additions will be endogenous if included in a demand equation for another commodity in the consumption set. In the fishing demand example, the boat ownership indicator is a dummy endogenous variable (Heckman 1978). This suggests that capital expenditure decisions should be modeled simultaneously with other aspects of the consumer problem.

The basic neoclassical model of consumption with durable goods has the consumer choosing the allocation of expenditures among nondurables and capital stocks to maximize intertemporal utility (Deaton and Muellbauer 1980). This model yields solutions for the optimal demands for nondurables and durable stock in each period that are functions of the existing durable stock, discounted prices and the user cost of capital for all periods over the planning horizon. With weak intertemporal separability, future prices are irrelevant to current decisions, so the expenditure for the nondurables x and the durable stock K in any period is a function of the existing durable stock K_{-1}, current period prices p, and the user cost of capital k

(3-1) $$e(p,u,b,s,\varepsilon,k,K_{-1}) \equiv px(p,u,b,s,\varepsilon,k,K_{-1}) + kK(p,u,b,s,\varepsilon,k,K_{-1})$$

where u is the utility level, b indicates the supply of petroleum rigs for fishing, s is a vector of observable control characteristics and ε represents unobservables.[2] The relevant nondurables and durables are those which are weak complements to (fishing at) the petroleum rigs.

The (compensated) demand system for this problem is found by differentiating the full expenditure function with respect to the price of variable goods and the capital stock

(3-2) $$\begin{aligned} \frac{\partial e}{\partial p_1} &= x(p,u,b,s,\varepsilon,k,K_{-1}) + k\frac{\partial K(p,u,b,s,\varepsilon,k,K_{-1})}{\partial p_1} \\ &\vdots \\ \frac{\partial e}{\partial p_N} &= x(p,u,b,s,\varepsilon,k,K_{-1}) + k\frac{\partial K(p,u,b,s,\varepsilon,k,K_{-1})}{\partial p_N} \end{aligned}$$

(3-3) $$\frac{\partial e}{\partial k} = p\frac{\partial x(p,u,b,s,\varepsilon,k,K_{-1})}{\partial k} + K(p,u,b,s,\varepsilon,k,K_{-1})$$

Following the discussion in Chapter 2, this simultaneous system can be used to estimate the value of changes access to the weakly complementary petroleum rigs. The compensating variation for a discrete change in fishing access to rigs from b^1 to b^0 is

[2] With intertemporal separability, the user cost of capital is simply the current cost of capital purchases. Furthermore, assuming no depreciation, K is actually a measure of additions to capital stock. This is seen by noting that capital stock changes in each period according to $K(p,u,b,s,\varepsilon,k,K_{-1}) = d(p,u,b,s,\varepsilon,k,K_{-1}) + (1-\delta)K_{-1}$ where d(·) is the demand for additions to capital stock and δ is the depreciation rate. If δ = 0, then changes in K are proportional to d.

$$(3\text{-}4)\quad \begin{aligned} CV^K(b^1,b^0) &= e(p,u^1,b^1,s,\varepsilon,k,K_{-1}) - e(p,u^1,b^0,s,\varepsilon,k,K_{-1}) \\ &= \left[px(p,u^1,b^1,s,\varepsilon,k,K_{-1}) + kK(p,u^1,b^1,s,\varepsilon,k,K_{-1})\right] \\ &\quad -\left[px(p,u^1,b^0,s,\varepsilon,k,K_{-1}) + kK(p,u^1,b^0,s,\varepsilon,k,K_{-1})\right] \end{aligned}$$

Note that to the extent capital stock is actual fixed over the decision period the second terms in (3-2) and the entire expression in (3-3) can be ignored in demand estimation. If capital stock is fixed before *and* after the change rig access, then the second terms in the last two lines of the CV in (3-4) can also be ignored. This is what is done in the conventional formulation of the travel cost model that focuses on the first term in (3-1) or the *variable expenditure function* (Conrad and Schroder 1991). The welfare measure in this case reduces to

$$(3\text{-}5)\quad CV^V(b^1,b^0) = px(p,u^1,b^1,s,\varepsilon,K_{-1}) - px(p,u^1,b^0,s,\varepsilon,K_{-1})$$

where since there is no longer a trade-off between spending on nondurables and spending on capital additions, the cost of capital is omitted. The structural demand approach defined below follows the conventional travel cost formulation to recover the variable cost welfare measure in (3-5). The alternative treatment effects approach presents a way to recover the welfare measure of the value of rig access in (3-4) that includes spending on fishing capital.

Structural Demand Approach

The structural approach to recovering the welfare measure in (3-4) requires estimation of at least part of the system of interdependent commodity and capital stock demands. Two strategies have been applied in the literature. The first strategy is to estimate a partial demand system with the commodities that are (assumed) interdependent with the public good (Shapiro and Smith 1981). The capital augmented model shown in system (3-2)-(3-3) also requires the estimation of equations for all of the interdependent capital stocks. The second and more common strategy is to aggregate the interdependent goods and estimate a demand equation or system for a composite commodity or system of composites. The travel cost model is a classic example of this second strategy where the composite commodity is 'trips' and the price index is 'travel costs.' This approach is taken for the present application. However, it is acknowledged that the approach is problematic in the travel cost model because the composite commodities are delineated according to activities and/or locations *chosen* by the household. See Chapter 2 for a discussion. Adding capital costs into activity-based price indices compound the problems because there is no straightforward way to allocate the fixed capital costs to any one activity type, location or specific trip (Pollak and Wachter 1975). Consequently, separate capital stock equations should be estimated simultaneously.

For the empirical application of the structural approach, I *do not* estimate a demand system with capital stock equations. I follow the conventional travel cost approach in assuming that capital stock is fixed over the decision period and estimate the demand for a trip-based composite commodity representing only system (3-2). However, the previous period capital stock *is* included among the influences of the trip demand decision. Trip demand is specified as a 'pooled-site' model for recreational fishing at petroleum rigs in the Gulf of Mexico.

Following the canonical discrete/continuous model in the Chapter 2, the structural trip demand system consists of an index equation and a trip demand equation

(3-6) $$R = \alpha_I^R + (P_r - P_{nr})\alpha_P^R + K_{-1}\alpha_{K_{-1}}^R + s\alpha_s^R + \varepsilon^R$$

(3-7) $$T_r = \alpha_I^{11} + P_r\alpha_r^{11} + P_o\alpha_o^{11} + K_{-1}\alpha_{K_{-1}}^{11} + s\alpha_s^{11} + \varepsilon^{11}$$

where R is a binary indicator that equals 1 if an individual fished at a rig in the previous year and zero otherwise[3], T_r is the total annual fishing trips to petroleum rigs, P_r is the average cost of a rig trip, P_o is the average cost of a non-rig trip, y is income, K_{-1} is the existing stock of fishing capital, s is a vector of socioeconomic control variables. The unobservables of the selection decision ε^R and trip demand ε^{11} are assumed to be joint normally distributed.

The number of rig trips per year in (3-7) is modeled as Poisson process (Hellerstein 1999) so that estimated trip demand equation is

(3-8) $$\text{Prob}\left[T_r = c\right] = \exp\left(-T_r\right) T_r^c / c! \qquad c = 0,1,2,\ldots..$$

The extra error term in the trip equation of (3-7) relaxes the usual Poisson assumption that the mean and variance of the estimator are equal. This allows for unobserved heterogeneity and addresses over-dispersion common in count models. The extra error term also allows a convenient way to model selectivity that parallels the standard approaches with linear models. An example application of this estimator is given in Haab and McConnell (2002) and the construction of the likelihood function is shown in Greene (1995 pp. 580-582).

Based on the discussion of demand interdependence in Chapter 2, the value of a public good change can be measured as areas behind the demand curves estimated with (3-6)-(3-8) before and after the change. However, the data used in this case study is from a cross-section before the public good change so the welfare effects have to be simulated.

---

[3] I assume that the net utility of the rig use decision can be modeled by a 'reduced form' index equation that is linear in variables. This simple approach is taken because of the complexity of the indirect utility function corresponding to the semi-log demand equation of the count model. With this assumption, only the unobservable portions of the rig use decision and the trip demand are related.

The simulation to recover value of fishing access to petroleum rigs involves an integration under the demand curve for rig trips from the current cost of a trip to a 'choke' cost. The expected annual consumer surplus consumer surplus of rig access is

(3-9)
$$
\begin{aligned}
E[S] &= \int_{\Psi} \left\{ \int_{P_r}^{\overline{P_r'}} \left[\phi(\varepsilon^{11}) T_r \right] dP_r \right\} d\varepsilon^{11} \\
&= \int_{P_r}^{\overline{P_r'}} \left\{ \int_{\Psi} \left[\phi(\varepsilon^{11}) T_r \right] d\varepsilon^{11} \right\} dP_r \\
&= -\frac{E[T_r]}{\alpha_r^{11}} \qquad \alpha_r^{11} \neq 0
\end{aligned}
$$

where the denominator is the coefficient on the rig trip cost variable from (3-7), $\phi(\cdot)$ is the normal probability distribution function, and Ψ is the support of the unobservables in the trip demand equation (Hellerstein 1999). Following the notation in Chapter 2, $\overline{\overline{P_r'}}$ is the choke price for the uncompensated demand equation. Based on Hanemann's (1980) derivation for the semi-log demand equation, the exact compensating variation for a loss in access to petroleum rigs for recreational fishing is

(3-10)
$$
\begin{aligned}
E[CV] &= \int_{\Psi} \left\{ \int_{P_r}^{\overline{P_r}} \left[\phi(\varepsilon^{11}) T_r \middle| u = u^1 \right] dP_r \right\} d\varepsilon^{11} \\
&= \int_{P_r}^{\overline{P_r}} \left\{ \int_{\Psi} \left[\phi(\varepsilon^{11}) T_r \middle| u = u^1 \right] d\varepsilon^{11} \right\} dP_r \\
&= \frac{1}{\alpha_y^{11}} \ln\left[1 - E[T_r] \frac{\alpha_y^{11}}{\alpha_r^{11}} \right] \qquad \alpha_y^{11} \neq 0,\ \alpha_r^{11} \neq 0,\ E[T_r] < \frac{\alpha_t^{11}}{\alpha_y^{11}}
\end{aligned}
$$

where $\overline{\overline{P_r}}$ is the choke price for the compensated demand equation. Note that reversal of integration implied in these measures requires that the travel price be *independent* of the unobserved heterogeneity. I use this simple measure, but note the assumption is questionable because as argued in Chapter 2, the travel price (index) is endogenously determined along with the trip quantity index (Haab and McConnell 1996).

The welfare measures use expected rig trips, adjusted for selectivity (Bockstael et al. 1990). I also correct the expected value of the dependant variable for the lognormal transformation implicit in the Poisson-Normal model of heterogeneity so expected rig trips are calculated as

$$(3\text{-}11)\quad E[T_r]=\frac{1}{N}\sum_{n=1}^{N}\left[\exp\left(\alpha_l^{11}+P_{r,n}\alpha_r^{11}+P_{o,n}\alpha_o^{11}+K_{-1,n}\alpha_{K_{-1}}^{11}+s_n\alpha_s^{11}+\sigma_{11R}\mu_n^{11}+\sigma_{11}^2/2\right)\right]$$

where $\mu_n^{11}=\phi\left(\hat{R}_n\right)/\Phi\left(\hat{R}_n\right)$ is the inverse Mills ratio term with $\Phi(\cdot)$ as the cumulative distribution function, σ_{11R} is the estimated covariance of the rig choice and rig trip decisions, and σ_{11} is the estimated standard deviation of unobservables in the rig trip equation. The last term is the correction of the mean for the lognormal distribution.[4] Note that I do not adjust the $E[S]$ and $E[CV]$ integration results for the appearance of the price and income variables in $\hat{R}_n$ of the expected trips equation. This adjustment is obviated by the assumption of exogenous prices that is used to reverse the integration in the welfare calculations. The model is estimated in LIMDEP (Greene 1995).

Treatment Effects Approach

The structural activity demand approach outlined earlier is problematic because the price index (travel costs) and quantity index (trips) will be endogenous to the consumer problem. Furthermore, although not included in the model specified for this chapter, the

[4] The model is estimated with the log of rig trips and the heterogeneity error term ε^{11} of this equation is assumed to be normal. Therefore, the trips within the Poisson probability are assumed to take a log-normal distribution. From Greene (2000) the expected value of the log-normal rig trips variable is $E[T_{r,n}] = exp(m + \sigma^2_{11}/2)$ where $m_n = E[Ln(T_{r,n})]$.

structural demand model requires a separate estimating equation to (simultaneously) incorporate the demand for capital additions.

An alternative strategy, introduced in Chapter 2, doesn't require separate price and quantity indices. Rather this *treatment effects* approach works to specify conditions whereby welfare changes can be measured directly by differences in observed spending by different segments of the population. For present case, I am suggesting that there is a hypothetical program to allow the use of petroleum rigs as artificial habitat for recreational fishing. Removal or expansion of this program can be considered a change in the supply of access to a public good.

Let b^1 be the reference condition of the rigs program and consider the special case analyzed in Chapter 2 where access to rigs is zero at b^0 so that $b^1 > 0$ and $b^0 = 0$. In this case, the recreational anglers who report fishing at the rigs are the program participants or treatment group and those who do not are the control group. The idea is to use the differences in expenditures between these two groups to evaluate the *with* and *without* program (petroleum rigs) welfare measure. The counterfactual assumptions required to uncover the value of access from the difference in spending with and without rig use are detailed in Chapter 2.

With the appropriate counterfactual assumptions, the *self-selection* decisions of anglers suggest three possible sources of differences in expenditures: 1) fishing at a rig may require higher (or lower) expenditures on average, 2) those who fish rigs may have an inherent tendency to spend more (or less) on fishing than those who do not fish rigs, and 3) the expenditures of those who fish rigs may change more (or less) because of a change in rig access than those who do not visit rigs, if they had. The various approaches to dealing with selection bias in treatment effects models all attempt to isolate the first effect by controlling or capturing the second two sources of variation from estimates (Vella and Verbeek 1999; Winship and Morgan 1999). However, as described in Chapter 2, the second two sources of spending differences provide important information about the relative value anglers place on access to fishing at rigs

As in the structural demand approach, the decision to fish rigs is modeled with a linear index equation

(3-12) $$D^1 = G\beta^{D^1} + \varepsilon^{D^1}$$

where $G = \{1, y, K_{-1}, s, z\}$, $\beta^{D^1} = \{\beta_I^{D^1}, \beta_y^{D^1}, \beta_{K_{-1}}^{D^1}, \beta_s^{D^1}, \beta_z^{D^1}\}$, and ε^{D^1} represents the unobservables at rig access level b^1. This index is motivated by the latent net utility of choosing to fish at least one petroleum rig in the previous year. All variables are defined as in the structural demand model and z is an exclusion restriction required so that we can manipulate an individual's probability of rig use without affecting their expenditures. Note that the latent net utility value can be different for individuals with the same observed characteristics because of the unobserved heterogeneity term ε^D. For example, some anglers won't fish at a rig unless they own a boat, while others will rent a boat or hire a charter to do so. This also suggests that capital purchases should be incorporated in expenditures and welfare measures as shown in (3-1) and (3-4).

The index equation defines an endogenous switching regime model of annual variable and capital expenditures with and without rig use

(3-13) $$[px + kK] = \begin{cases} X\beta^{11} + \varepsilon^{11} & \text{if } D = 1 \\ X\beta^{01} + \varepsilon^{01} & \text{otherwise} \end{cases}$$

where $X = \{1, y, K_{-1}, s\}, \beta^{ij} = \{\beta_I^{ij}, \beta_y^{ij}, \beta_{K_{-1}}^{ij}, \beta_s^{ij}\}$ and ε^{ij} represents the unobservables in the expenditure equations at rig access level *j*.[5] Presently, $j = 1$ because b^1 is used as the reference state of rig access. The superscript *i* corresponds with the binary indicator *D* and equals 1 if rigs were used and zero otherwise.

Following Phlips (1983), income is normalized to the own price level and the intercept and error term for the public good users ($i = D = 1$) are implicitly defined as $\beta_I^{11} = \tilde{\beta}_I^{11} p^{11} + \tilde{\beta}_p^{11} p^{01}$ and $\varepsilon^{11} p^{11}$, respectively. These terms are defined similarly for the nonusers ($i = D = 0$). Therefore, prices appear endogenously as a portion of the unobservable determinants of spending.[6] That is, *p* is not formally defined in terms of a price index separate from total expenditures. The treatment effects approach can model prices this way because they are not required in the derivation of welfare measures. This is useful for reasons discussed in the introduction, especially when dealing with cross-sectional data where price variation is commonly an expression of changes in some other variable (e.g., distance or quality). The added flexibility is also particularly important when the public good of interest is defined as a characteristic of a nonmarket activity (Bockstael and McConnell 1993). In such cases, including the current study, it is easier to identify the expenditures on the nonmarket activity (e.g., fishing), than to define the

---

[5] I experimented with other functional forms for the expenditure equations such as the quadratic in income specification consistent with the quadratic almost ideal demand system. However, as frequently occurs with recreational expenditure data, the income terms did not come up significant in any specification. Therefore, I opted for the simpler linear specification with the additive error term. It is easier to derive a linear in variables net utility function from the linear Engel equation.

[6] This specification can be integrated to recover the related indirect utility function (Hausman 1981) that can be used to specify the form of the related net utility index. In the Appendix I show that the resulting net utility index can be reduced to a simple linear in variables equation with an additive error.

price and quantity indices necessary to estimate a structural demand model for the activity.

The full endogenous switching estimating system is

(3-14) $$D^{1} = G\beta^{D^1} + \varepsilon^{D^1}$$

(3-15) $$[px + kK]^{11} = X\beta^{11} + \varepsilon^{11}$$

(3-16) $$[px + kK]^{01} = X\beta^{01} + \varepsilon^{01}$$

Assuming ε^{D^1}, ε^{11}, and ε^{01} are joint normally distributed, the parameters $\{\beta^{11}, \beta^{01}, \beta^{D^1}, \sigma^{11}, \sigma^{01}, \rho^{11D^1}, \rho^{01D^1}\}$ can be estimated simultaneously via maximum likelihood or in a two step procedure for simultaneous equations with endogenous switching (Maddala 1983).[7] I obtain FIML estimates of the model parameters using the endogenous switching estimator in LIMDEP (Greene 1995).

As defined in Chapter 2, the standard treatment effect and the policy relevant treatment effect measures for this model are (Heckman, Tobias and Vytlacil 2001):

(3-17) $$ATE^{\hat{\Delta}^P} = E\left[\hat{\Delta}^P \middle| y,s\right] = X\left(\beta^{11} - \beta^{01}\right)$$

(3-18) $$TT^{\hat{\Delta}^P} = E\left[\hat{\Delta}^P \middle| y,s,D=1\right] = X\left(\beta^{11} - \beta^{01}\right) + \left(\sigma^{11D} - \sigma^{01D}\right)\lambda^{11}$$

(3-19) $$TU^{\hat{\Delta}^P} = E\left[\hat{\Delta}^P \middle| y,s,D=0\right] = X\left(\beta^{11} - \beta^{01}\right) + \left(\sigma^{11D} - \sigma^{01D}\right)\lambda^{01}$$

(3-20) $$E\left[\hat{\Delta}^{P^*}\right] = X\left(\beta^{11} - \beta^{01}\right) + \left(\sigma^{11D^1} - \sigma^{01D^1}\right)\left[-G\beta^{D^1}\right]$$

---

[7] The covariances are easily recovered from the correlation coefficient because the variance of the index equation is normalized to unity.

$\lambda^{11} = \phi\left(G\hat{\beta}^{D'}\right) / \Phi\left(G\hat{\beta}^{D'}\right)$, and $\lambda^{01} = -\phi\left(G\hat{\beta}^{D'}\right) / \left[1 - \Phi\left(G\hat{\beta}^{D'}\right)\right]$. Heckman, Tobias, and Vytlacil (2001) show simple unconditional estimators for each of the four treatment effect parameters as

(3-21) $$E[K] \approx \frac{1}{N} \sum_{n}^{N} K(X_n, z_n)$$

where K is the treatment effect measure of interest and N is the number of observations in the relevant set, i.e., N is the whole sample for (3-17) and (3-20), only the users for (3-18) and only the nonusers for (3-19). Note that expectation in (3-20) can be conditioned on any subset of the sample. For example, evaluating (3-20) over the set of rig users, gives the expected treatment effect welfare measure for a randomly chosen individual from this group. This calculation and a similar one for the group of nonusers is reported in the results.

Data

The sample for the analysis is taken from the 1999 U.S. National Marine Fisheries Service Economic Survey of Private Boat Anglers. A subset was selected from the sample of recreational anglers along the Gulf of Mexico coast (Alabama to Texas). See QuannTech (2001) for more information about the intercept and phone survey instruments and the dataset. The surveys elicited detailed information about fishing location, target species, and expenditures for variable and capital goods. In particular, respondents were asked to report the number of days that they fished "within 300 feet of an oil or gas rig or within 300 feet of an artificial reef created from an oil or gas rig" during the prior year. This information allowed the sample to be split into a segment that fished at rig sites during the previous year and all other respondents. Brief descriptions of the coding of the relevant variables appear in Table 3.

Note that nearly half the sample did not report household income. Missing income values were replaced with the mean reported value from the respondent's county of residence. There was also missing trip cost data. The portion of the sample who took

Table 3. Sample means and standard deviations for rigs model variables

Variables	Users (n=372)	Nonusers (n=124)	All (n=496)
Cost of a Rig Trip ($)	89.978	96.932	92.375
	(52.181)	(18.280)	(43.685)
Cost of a Non-rig Trip ($)	153.470	52.435	118.637
	(455.607)	(64.946)	(373.736)
Cost Difference for Rig Trip ($)	-63.492	44.497	-26.262
	(456.379)	(63.740)	(374.720)
Total Fishing Trips	29.628	25.514	28.210
	(31.769)	(42.623)	(35.878)
Rig Fishing Trips	14.241	0.000	9.332
	(19.572)	(0.000)	(17.226)
Total Variable Expenditures ($)	3,406.087	1,351.461	2,697.739
	(7,427.224)	(3,734.135)	(6,469.534)
Total Annual Expenditure ($)	7,602.246	2,512.929	5,847.666
	(14,137.100)	(4,854.344)	(12,033.485)
Current Capital Stock ($)	11,138.121	5,045.676	9,037.705
	(17,584.240)	(7,352.223)	(15,145.725)
Capital Stock Lagged ($)	6,941.962	3,884.208	5,887.778
	(14,078.927)	(6,859.418)	(12,166.827)

Standard deviations in parentheses. Table continued below.

Table 3. Sample means and standard deviations for rigs model variables (cont.)

Variables	Users (n=372)	Nonusers (n=124)	All (n=496)
Income ($/yr)	65,863.024	60,782.662	64,111.531
	(34,709.303)	(23,628.382)	(31,400.900)
Experience (years)	23.994	22.077	23.333
	(14.314)	(16.159)	(14.983)
Gender (1=female)	0.082	0.113	0.092
	(0.274)	(0.318)	(0.290)
Memberships (1=yes)	0.184	0.113	0.160
	(0.388)	(0.318)	(0.367)
Louisiana Resident (1=yes)	0.322	0.444	0.364
	(0.468)	(0.499)	(0.482)
Mississippi Resident (1=yes)	0.093	0.162	0.117
	(0.290)	(0.370)	(0.321)
Texas Resident (1=yes)	0.231	0.225	0.229
	(0.422)	(0.419)	(0.420)
Coastal Resident (1=yes)	0.919	0.911	0.916
	(0.273)	(0.287)	(0.277)
Target Rig Species (1=yes)	0.262	0.062	0.193
	(0.440)	(0.241)	(0.395)

Standard deviations in parentheses.

both rig and non-rig trips had missing data because the intercept data only reflects one of these type of trips. Similarly, those who did not take a rig trip had no expenditure data for this type of activity. The missing trip cost values were replaced with the mean values over only rig users in the relevant Gulf State in order to avoid mixing across the rig and non-rig groups. The replacement procedure is summarized in Table 4.

Table 4. Replacement rules for missing variable cost data

		At least one rig trip in the previous 12 months?	
		YES (1)	NO (0)
Took a rig trip when intercepted?	YES (1)	Cost of a rig trip = VC_{11} Cost of non-rig trip = $\overline{VC}_{01}^{state}$ (218)	Not Possible (0)
	NO (0)	Cost of a rig trip = $\overline{VC}_{11}^{state}$ Cost of non-rig trip = VC_{01} (154)	Cost of a rig trip = $\overline{VC}_{11}^{state}$ Cost of non-rig trip = VC_{00} (124)

The summary statistics in Table 3 are split into two sub-samples: those anglers who fished at rigs (users) in the previous year and those who did not (nonusers). The socioeconomic characteristics reported are fairly consistent across the sample. The key differences between the two sub-samples arise with respect to the economic decision variables such as trip costs, expenditures, capital stock holdings, and rig species targeting. Specifically, a rig trip costs relatively more than a non-rig trip for nonusers. The converse is true for users suggesting that each group has an absolute advantage in their chosen activity.[8] Each group also appears to have a comparative advantage in their chosen activity. However, cost savings per trip is only part of the story. Advantages cannot be fully studied without reference to each group's willingness to pay for rig and non-rig fishing. This premise is explored in the results, although, the differences in annual fishing expenditures and capital stock of the two groups in Table 3 is suggestive.

[8] The terms 'absolute' and 'comparative advantage' are commonly used in the labor supply literature. For example, when based on earnings, either advantage can be used to explain the type, variety, or location the of labor selected by an individual (Emerson 1989; Maddala 1983). In the case of recreational angling, the advantages are measured in terms of (utility constant) cost savings for different types or locations of fishing.

The pooled site travel cost model is estimated using the number of annual fishing trips to rigs as the dependent variable. Independent variables in this regression include the variable cost of a rig trip, the variable cost of a non-rig trip, the stock of fishing durables at the beginning of the year (capital stock lagged), household income and other control variables. The cost of rig and non-rig trips is defined as any personal spending by the respondent for the trip on which they were intercepted. A list of spending categories included in variable trip costs is shown in Table 5. Note that the opportunity cost of time is not included among the variable cost items. Time costs are not considered in the present analysis.

Table 5. Spending included in the variable and capital fishing expenditures

Variable	Capital
travel	boat
lodging	motor
food	trailer
drink	electronics
boat fuel	safety gear
boat rental	rigging for fishing
dock fees	boat or equipment repairs
launch fees	rods and reels
repairs and towing on trip	fishing line
bait	lures and artificial bait
special licenses for trip	other fishing equipment
tackle and guide services	fishing books and magazines
equipment rental	fishing club memberships
special clothing	camping equipment for fishing
film, sundries, and souvenirs	fishing licenses

Two treatment effects models are estimated as defined in (3-14)-(3-16). The first uses annual variable trip rig and non-rig trip expenditures as the dependent variable in the outcome equations in the treatment effects switching regression. These variables are calculated by multiplying the respondent's rig (non-rig) trip expenditures by their total number of annual rig (non-rig) trips. The second treatment effects model uses *total annual expenditures* calculated by adding the additions to capital stock during the year to the annual variable expenditures. Spending categories included in the capital measures are listed in Table 5.

The other control variables, as well as the variables used in the rig use selection equation, are listed in the results. One variable of note, however, is the decision to target rigs species. This variable is coded 1 if the respondent indicated a target preference for species that are commonly associated with oil and gas rig habitat. The means for the target variable indicate that a larger portion of those who fish at rigs also target rig species. This introduces the potential modeling issues associated with multiple criteria for selectivity (Maddala 1983 pp. 278-283). In the recreational fishing demand literature the relevant questions concern, for example, whether anglers choose a species target and fishing location sequentially or simultaneously and, if sequentially, in what order (Kling and Thomson 1996). I will assume that the process is sequential and the target decision is made first by using a binary target variable as a regressor in the rigs decision equation. The target variable also serves as the exclusion restriction necessary for the index function set-up (Heckman and Vytlacil 2001a).

Results

Travel Cost Model

The travel cost count model estimation results are shown in Table 6. Three variables are significant in the rig use decision equation. Mississippi residents are less likely to fish at rigs compared to anglers from other states and those who target species associated with rigs are more likely to fish rigs. The negative coefficient on the 'cost difference' variable suggests that higher relative rig costs decrease the probability of fishing rigs.

A number of the parameters in the trip demand equation are significant. Those with higher capital stock at the beginning of the year tend to take relatively more rig trips. Those living near the coast, residents of Mississippi and Texas, and members of fishing clubs are also take relatively more rig trips.[9] Females, Louisiana residents, and individuals with more experience take less rig trips. Those with higher income also appear to take less rig trips, suggesting a negative income effect. However, given the measurement problems with the income variable, this result is not especially troubling.

The own price variable (cost of a rig trip) is negative, but not significant, whereas, the substitute price term (cost of a non-rig trip) is significant and positive. The latter result implies that rig and non-rig fishing trips are substitutes. Although not significant, the inverse of the own price coefficient gives an expected consumer surplus per trip of $4,442. See equation (3-9). Multiplying this value by the expected trips as shown in Table 7 gives an expected annual uncompensated surplus of rigs fishing of $37,824. Adjusting for income effects using expression (3-10), the corresponding expected annual compensating variation is lower, but still very high, at $27,569. The standard deviations shown in the table were obtained by evaluating the measures for each individual in the sample.

The variance of the extra stochastic term in the Poisson-Normal model is significant, indicating that there is unobserved heterogeneity influencing the trip decision.

---

[9] Residents of Alabama are the base case when all other State dummy variables are equal to zero.

Table 6. Estimates for the Poisson-normal travel cost model with selectivity

Variables	Rigs Decision	Trip Demand
Constant	2.63E-01	1.49E+00
	(3.12E-01)	(1.97E-01)*
Cost of a Rig Trip ($)		-2.25E-04
		(2.91E-04)
Cost of a Non-rig Trip ($)		2.25E-04
		(7.53E-05)*
Cost Difference for Rig Trip ($)	-3.86E-03	
	(8.91E-04)*	
Capital Stock Lagged ($)	1.10E-05	1.67E-05
	(8.26E-06)	(1.45E-06)*
Income ($/yr)	3.86E-07	-3.42E-06
	(2.52E-06)	(6.20E-07)*
Experience (years)	4.64E-03	-7.84E-03
	(4.19E-03)	(1.91E-03)*
Gender (1=female)	-1.69E-01	-2.83E-01
	(2.13E-01)	(1.23E-01)*
Memberships (1=yes)	2.93E-01	5.20E-01
	(1.79E-01)	(5.25E-02)*
Louisiana Resident (1=yes)	-1.08E-01	-2.66E-01
	(1.95E-01)	(5.71E-02)*
Mississippi Resident (1=yes)	-5.07E-01	2.32E-01
	(2.84E-01)*	(6.69E-02)*
Texas Resident (1=yes)	-5.22E-02	1.60E-01
	(2.15E-01)	(6.24E-02)*
Coastal Resident (1=yes)	8.37E-02	6.19E-01
	(2.08E-01)	(1.77E-01)*
Target Rig Species (1=yes)	7.64E-01	
	(2.73E-01)*	
σ^{trips}		9.55E-01
		(3.08E-02)*
$\rho^{selection,trips}$		5.46E-01
		(2.59E-01)*
$\sigma^{selection,trips}$		5.22E-01
		(2.59E-01)*

Standard errors are shown in the parentheses below each estimate.

*Estimate significant at the 0.10 level.

The final value of the log likelihood function is -1603.968.

In addition, the correlation and covariance between the rig use and trip count decisions are significant suggesting that selectivity is present as modeled.

Table 7. Count model welfare analysis for loss of rigs access

Mean Actual Rig Trips	14.24
	(19.57)
Expected Rig Trips	8.51
	(7.38)
Expected Annual Compensating Variation	27,569
	(12,850)
Expected Annual Consumer Surplus	37,824
	(32,784)

Standard deviations shown below the means.

Treatment Effects Models

The FIML estimated coefficients of the treatment effects model (TEM) with annual variable expenditures and total annual expenditures are shown, respectively in Table 8 and Table 9. The signs and levels of the significant coefficients in the rigs decision equations are roughly consistent with those estimated in the selection equation of the travel cost model (TCM).[10] Again, those who target rig species are more likely to fish at rigs and the level of existing capital stock is not a significant influence on the probability of fishing at rigs. Mississippi residents are less likely to fish rigs than Texas and

[10] The coefficients between TCM count demand estimates and the TEM users expenditure equation are not directly comparable because they each measure influence on a different dependant variable.

Table 8. Annual variable expenditures treatment effects model results

Variables	Rigs Decision	With Rig Use	Without Rig Use
Constant	1.21E-01	3.07E+03	4.57E+03
	(3.83E-01)	(2.52E+03)	(9.31E+03)
Capital Stock Lagged ($)	1.45E-05	8.37E-02	4.58E-02
	(9.99E-06)	(2.73E-02)*	(1.18E-01)
Income ($/yr)	1.99E-06	-8.59E-03	-1.30E-02
	(2.67E-06)	(1.74E-02)	(3.05E-02)
Experience (years)	5.52E-03	-5.64E+01	-1.99E+01
	(4.47E-03)	(4.38E+01)	(5.12E+01)
Gender (1=female)	-1.71E-01	-2.16E+03	-1.71E+03
	(2.18E-01)	(2.35E+03)	(2.52E+03)
Memberships (1=yes)	2.63E-01	2.36E+03	7.58E+02
	(1.94E-01)	(1.33E+03)*	(2.55E+03)
Louisiana Resident (1=yes)	-3.99E-01	-4.18E+01	2.17E+02
	(1.95E-01)*	(2.33E+03)	(3.71E+03)
Mississippi Resident (1=yes)	-6.35E-01	1.13E+04	-1.65E+02
	(2.78E-01)*	(1.76E+03)*	(4.83E+03)
Texas Resident (1=yes)	-2.69E-01	1.31E+03	1.92E+03
	(2.35E-01)	(2.18E+03)	(2.96E+03)
Coastal Resident (1=yes)	1.08E-01	2.54E+02	-2.21E+03
	(3.26E-01)	(1.55E+03)	(1.55E+03)
Target Rig Species (1=yes)	8.13E-01		
	(2.78E-01)*		
σ^{expend}		6.50E+03	3.56E+03
		(1.78E+02)*	(9.81E+02)*
$\rho^{selection,expend}$		-4.73E-02	1.59E-01
		(3.89E-01)	(2.29E+00)
$\sigma^{selection,expend}$		5.66E+02	-4.25E+06
		(8.31E+03)	(6.77E+06)

Standard errors are shown in the parentheses below each estimate.

*Estimate significant at the 0.10 level.

The final value of the log likelihood function is -5244.391.

Alabama residents. The same is true for Louisiana residents which is somewhat surprising given that the majority of petroleum platforms are located off the coast of Louisiana.

Only five of the estimates in the annual variable expenditure outcome equations are appreciably significant and most of these coefficients are for the 'Rig Use' equation. A similar pattern appears for the total annual expenditure outcome equations. Interestingly, the existing level of fishing capital has a positive influence on annual variable expenditures, but not on total annual expenditures. Based on the TCM results in Table 6, the additional spending arises because those with larger fishing capital stocks take relatively more rig trips. However, it appears that these individuals are not any more likely to add to capital stock throughout the year than those with relatively smaller capital stocks. Those with paid memberships to fishing clubs and residents of Mississippi tend to spend more on variable and capital costs for rig trips.

The estimated variances of all the spending outcome equations in Table 8 and Table 9 are significant, indicating the importance of unobserved heterogeneity in this sample. However, because of relatively insignificant correlations, the covariances between the rigs decision and spending equations are not significant. This suggests that there is a limited degree of self-selection based on rig use in the sample. To use the analogy from the labor literature (Emerson 1989), the lack of significant covariance between the use and spending decisions implies that neither group has an 'absolute advantage' in their selected option. In the present application, an individual has an absolute advantage in their chosen activity if that activity offers them a significantly lower cost for utility than competing activities. For example, those with an absolute advantage for rig use can

Table 9. Total annual expenditures treatment effects model results

Variables	Rigs Decision	With Rig Use	Without Rig Use
Constant	1.20E-01	1.09E+04	6.51E+03
	(3.61E-01)	(4.59E+03)*	(6.34E+03)
Capital Stock Lagged ($)	1.44E-05	-1.60E-02	8.18E-03
	(8.96E-06)	(6.42E-02)	(8.37E-02)
Income ($/yr)	2.06E-06	2.00E-02	-1.28E-02
	(2.66E-06)	(2.49E-02)	(3.48E-02)
Experience (years)	5.45E-03	-9.49E+01	-2.31E+01
	(4.59E-03)	(6.98E+01)	(5.17E+01)
Gender (1=female)	-1.59E-01	-3.81E+02	-2.48E+03
	(2.20E-01)	(3.24E+03)	(2.21E+03)
Memberships (1=yes)	2.60E-01	4.55E+03	3.21E+03
	(2.16E-01)	(2.25E+03)*	(2.19E+03)
Louisiana Resident (1=yes)	-3.94E-01	-2.32E+03	-1.73E+03
	(1.92E-01)*	(3.31E+03)	(2.35E+03)
Mississippi Resident (1=yes)	-6.31E-01	2.26E+04	-1.51E+03
	(2.69E-01)*	(2.38E+03)*	(3.09E+03)
Texas Resident (1=yes)	-2.68E-01	-1.61E+03	6.60E+02
	(2.25E-01)	(2.95E+03)	(2.51E+03)
Coastal Resident (1=yes)	1.02E-01	-3.64E+03	-1.21E+03
	(2.80E-01)	(3.15E+03)	(1.89E+03)
Target Rig Species (1=yes)	8.24E-01		
	(2.51E-01)*		
σ^{expend}		1.22E+04	4.66E+03
		(4.43E+02)*	(6.15E+02)*
$\rho^{selection,expend}$		-1.08E-01	1.88E-01
		(3.29E-01)	(9.01E-01)
$\sigma^{selection,expend}$		-1.32E+03	8.78E+02
		(4.05E+03)	(4.30E+03)

Standard errors are shown in the parentheses below each estimate.

*Estimate significant at the 0.10 level.

The final value of the log likelihood function is -5494.606.

can attain the same level utility at a lower cost by using rigs than by not fishing at rigs. A similar condition applies for those who chose not to use rigs. The weaker condition of 'comparative advantage' implies that the average user (nonuser) spends less for the same utility level than the average nonusers (user) when they both (do not) use rigs. A null hypothesis of no comparative advantage can be evaluated with a joint test of $\beta^{11} = \beta^{01}$ and $\sigma^{11D} = \sigma^{01D}$. The Wald statistics of 59.21 and 135.06 for these restrictions in the variable and total expenditures models, respectively, rejects joint equality with greater than 99% confidence. Thus, there is still significant information in the rig use (self-selection) decisions of anglers in the sample that can be used to evaluate the relative valuations of rig access. The treatment effect welfare measures introduced in Chapter 2 are designed to exploit this information.

The unconditional treatment effects and welfare measures of rig access are shown in Table 10. These figures are obtained by evaluating expressions (3-17) through (3-20) for each individual in the relevant group and averaging as defined in (3-21). Note that even the largest measure shown in this table is still less than a third of the values shown in Table 7 for the travel cost model. The relatively large welfare measures in the travel cost model are due primarily to the small coefficient estimated on the travel cost parameter. As described in Chapter 2, the price variable is crucial in welfare analysis with the travel cost model. The treatment effects model sidesteps this reliance by using information from all of the model coefficients to generate welfare measures. However, more research is need on ways to analytically and empirically compare the travel cost and treatment effects approaches.

Table 10. Annual expenditure treatment effects and welfare estimates of rig access

Parameter	TEA Variable Costs	TEA Total Costs
ATE	1,737	4,950
	(3,906)	(8,131)
E[CV]	2,141	5,968
	(3,861)	(7,926)
TT	1,101	3,264
	(3,570)	(7,276)
E[CV | users]	2,054	5,659
	(3,591)	(7,268)
UT	2,945	8,156
	(4,377)	(9,158)
E[CV | nonusers]	2,308	6,556
	(4,339)	(9,047)

Standard deviations shown below the estimates.

The average treatment effect and the policy relevant measure of compensating variation of lost rig access for the whole sample are shown in the first pane. A randomly selected angler will spend an additional $1,737 in variable costs annually to fish rigs. This amount more than doubles to $4,950 when expenditures on fishing capital is included. The randomly chosen individual is willing to pay between $2,141 and $5,968 annually for access to rigs for fishing where the upper end measure includes forgone capital spending. A randomly chosen angler from the group that used rigs is willing to pay between $2,045 and $5,659 annually and has an expected annual cost for rig fishing of between $1,101 and $3,264. Similarly, a randomly chosen angler from the group anglers who did not use rigs is willing to pay between $2,308 and $6,556 annually and has an expected annual cost for rig fishing of between $2,945 and $8,156.

The results in Table 10 are in line with model checks proposed in Chapter 2. For users the expected compensating variation of use is considerably greater than the spending difference measure. This is consistent with a preference for rig use, but it also implies that users have a lot to loose if they are denied access to rigs. The nonuser results show an expected compensating variation that is less than the extra annual costs of rig use. This explains way this group does not fish at rigs. Interestingly, though, a randomly selected nonuser actually has a relatively higher value for rig access than a randomly chosen user or a randomly chosen individual in the sample. This is true of the variable cost and total cost results. Thus, nonusers value rig access, but do not fish at rigs because doing so requires a relatively more expenditure, especially when additions to fishing capital are considered. These results illustrate that nonusers do not use rigs because the benefits are less than the cost. As for the rest of the sample, the results in the first pane of Table 10 indicate that a randomly selected individual will choose to use rigs because the benefits do outweigh the costs. The statistical significance of these measures was not tested directly, but the rejection of joint equality in the comparative advantage test is suggestive. Note also that, although not shown, all of the consistency tests are met at the minimum and maximum values of the sample.

Discussion

This chapter has explored the role of capital expenditures in revealed preference modeling of recreation decisions. The treatment effects approach developed in Chapter 2 was used to evaluate the welfare effects of restricting access to fishing at petroleum rigs in the Gulf of Mexico. A travel cost trip demand model was also estimated.

Based on the treatment effects model, the artificial fishing habitat offered by petroleum rigs was found to 'cause' a \$1,737 to \$4,950 increase in average annual fishing expenditures among anglers in the sample. The amount that a randomly chosen angler would be just as well-off without rigs is estimated ranges from $2,141 to $5,968. The upper end of the range is the welfare effect including the additions to fishing capital. These estimated values are substantially lower than the welfare measures generated with the travel cost model. This is peculiar result could be because of the sensitivity of travel cost welfare measures to when cost-based prices are used (English and Bowker 1996; Wilman and Pauls 1987).

The variation in the treatment effects model measures suggests that not considering information about recreation capital acquisitions and holdings could seriously understate the opportunity cost of restricting access to fishing at rig habitat. The results suggest a need to consider ways to incorporate recreation capital in other revealed preference valuation exercises.

CHAPTER 4
APPLICATION TO PUBLIC UTILITY PRICING

If a rational consumer does not know the price of a purchased commodity, then he cannot optimally adjust budget allocations and marginal valuations to be in line with that price.[1] When prices accurately reflect social opportunity costs (as in a perfectly competitive market), the burden of price misperception is on the consumer. That is, following Shin's (1985) hypothesis, the consumer accepts the inefficiencies from price misperception in return for the avoided cost of determining the actual price. On the other hand, when prices *do not* accurately reflect social opportunity costs, then price misperception and the *sub-optimal* consumption levels have welfare implications beyond the consumer's budget allocations. Classic cases of deviations from socially optimal prices can occur, for example, in the presence of externalities and/or in the context of administered prices for monopoly services (Carter and Milon 1998).

This chapter develops analytical and empirical models to evaluate price misperception and the value of price information for the case of administered prices for public utility service. In doing so, the focus is on the consumer's self-reported *price awareness*. This perspective is more fundamental than the studies that have used data across utilities to examine the effect of different levels of information provision and rate structure style on the quantity of public utility service demanded. See Cavanagh Hanemann and Stavins (2001) for a review. The perspective is closer to the large body of

---

[1] In the case considered here the consumer can *choose* not to know the price even though this information is available with certainty. This is different than the case of choice under *irreducible* price uncertainty (Johnansson 1991).

research debating the 'correct' price specification in models of demand in the presence of nonlinear budget constraints (i.e., block pricing).[2] The portions of this research that have attempted to empirically test a consumer's perception of the price of service is of central interest (Chicoine and Ramamurthy 1986; Griffin and Chang 1990; Nieswiadomy 1992; Nieswiadomy and Molina 1991; Shin 1985). This chapter extends this research by providing a theoretical framework to analyze the comparative statics of price misperception and identifies the value of complete price knowledge. Importantly, this framework considers the possibility that price elasticity of demand may change when price perception changes because of changes in exogenously supplied price information.[3]

The theoretical framework is used to develop a structural model of public utility demand and a treatment effects model of expenditures on utility services. Both models are based on the discussion of structural and treatment effects approaches to measuring public good values in Chapter 2. Exogenously supplied information about the price of utility service is the public good in this case. The results from the structural and treatment effects models are used to evaluate the benefits of a hypothetical program that would fully inform customers about the price of service.

Price Perception and the Value of Price Information

The welfare implications of price misperception with administered prices depend on the goals of the pricing authority (e.g., public utility) and the relative abilities of the authority and the consumer to accurately gauge the social opportunity cost of

[2] Witness the lively exchanges in *Land Economics* among Foster and Beattie, Griffin et. al., Opaluch, Charney and Woodard, Billings and Agthe, Ohsfeld, Polzin and Stevens et al. in the early 1980's and subsequent literature that continued throughout decade in that journal and *Water Resources Research*.
[3] Shin (1985) considers the value of price information for the case where the consumer overestimates the actual price. However, his representation implicitly assumes that the response to price changes is the same regardless of the consumer's information set.

consumption. If the pricing authority aims to set prices to reflect marginal opportunity costs and/or they have explicit conservation goals, then they should also be concerned that these prices are perceived accurately. Furthermore, when the authority's administered price reflects the social opportunity cost of service more accurately than the consumer's perceived price, the authority can use price information as a policy tool. This follows because the relative costliness of a consumer's price information is partly a function of the amount of exogenously supplied information (e.g., advertising). The exogenously supplied price information is not itself sold in the market because, like advertising in general, it has public good characteristics that may favor other indirect financing schemes (Frech 1979).

Consider an consumer who perceives the price of a commodity Q as a function $\hat{p}(p,b)$ of the actual price p and an *exogenous* information supply b such as advertising or billing inserts that is available to all consumers.[4] It is assumed that b is a weak complement of Q so that the consumer is indifferent to the supply of b when Q is not purchased (Maler 1974). This implies that b is not a direct source of utility (i.e., b has no nonuse value) and, therefore, does not appear as a separate argument in the indirect utility function. Note that the distinction between $\hat{p}$ and p roughly corresponds to Pollak's (1977) conception of *normal* and *market* prices. In this case, $\hat{p}$ is a normal price signal that affects choices, whereas p is the market price that enters the budget constraint. This situation is exceedingly complicated to represent as a direct utility maximization problem

---

[4] As long as the consumers are price takers (e.g., subject to administered prices) the analysis can continue in a partial equilibrium framework. In other cases, however, a general equilibrium treatment is required because the equilibrium price will be an (inverse) function $p(\hat{\mathbf{p}},b)$ of the existing perceived price(s).

so only indirect utility and expenditure function representations of preferences are considered.

An accurate price perception is costly to the consumer so that $c(b)\geq 0$, but this cost decreases with the level of the exogenous price information provided, $\partial c(b)/\partial b < 0$.[5] The conditional indirect utility function for this problem is

$$v\left(\hat{p}(p,b), y - c(b), s, \varepsilon\right) \tag{4-1}$$

where s is a vector of observed individual control characteristics and ε summarizes the unobserved individual preferences and characteristics. The price index of a numeraire commodity is normalized to unity and is suppressed along with individual specific subscripts to simplify notation.

The marginal value of a change in the supply price information has two effects:

$$\frac{dy}{db} = \frac{\partial c}{\partial b} - \frac{(\partial v/\partial \hat{p})(\partial \hat{p}/\partial b)}{\partial v/\partial (y - c(b))} \tag{4-2}$$

where the first term on the right hand side is the reduction in the cost of an accurate price perception and the second term is the value of the change in price perception induced by the additional information.

To carry the analysis a step further, assume that the perceived price function takes the form $\hat{p}(p,\hat{b}) = p\cdot\hat{b}$ for $\hat{b} = (0, \infty]$. The perceived price is assumed proportional to the actual price by an adjustment factor $\hat{b}$ that summarizes the stock of exogenous price information. The perceived price is bound below to be greater than zero because the

[5] There have been attempts to explicitly consider the costs of price information search (Kolodinsky 1990). The more compact representation suffices for present purposes.

consumer will likely recognize a non-zero price if they are spending income on the commodity.

There are three cases to consider: (a) $\hat{b} = 1$, (b) $\hat{b} > 1$, and (c) $0 < \hat{b} < 1$. In the first case, the perceived price equals the actual price, i.e., the consumer is fully informed about price. In terms of the expression (4-2) above, the first case suggests that the additional information will not affect those who already know the price, beyond the reduction the cost of additional price information. The second case indicates a perceived price that is greater than the actual price and the third case indicates a perceived price that is less than the actual price. In the latter two cases the price misperception is leading to over or under consumption of the commodity Q relative to the composite commodity. This is most directly shown by differentiating the expenditure function for this problem by the *actual* price of Q

$$\text{(4-3)} \qquad \frac{\partial e\left(\hat{p}(p,b),u,s,\varepsilon\right)}{\partial p} = \frac{\partial e}{\partial \hat{p}}\frac{\partial \hat{p}}{\partial p}.$$

where $e(\cdot)$ gives the minimum expenditure required to attain utility level u. Noting that (4-3) is an expression of the compensated demand Q^c for Q we can rearrange to get

$$\text{(4-4)} \qquad \begin{aligned} \frac{\partial \hat{p}}{\partial p}\frac{\partial e}{\partial \hat{p}} &= Q^c\left(\hat{p}(p,b),u,s,\varepsilon\right) \\ \frac{\partial e}{\partial \hat{p}} &= \frac{Q^c\left(\hat{p}(p,b),u,s,\varepsilon\right)}{\partial \hat{p}/\partial p} \\ &= \frac{Q^c\left(\hat{p}(p,b),u,s,\varepsilon\right)}{\hat{b}} \quad \textit{for} \quad \hat{p}(p,b) = p\cdot\hat{b}. \end{aligned}$$

Thus, in the second (third) case where the perceived price is above (below) the actual price, the efficient compensated demand will be deflated (inflated) by $\partial\hat{p}/\partial p$ or $\hat{b}$ so that the consumer will be under (over) consuming Q.

To consider the value of a discrete change in price information it will be useful to explicitly define the conditional indirect utility function in terms of the decision to know the price

(4-5) $$v^{ij} = v\left(i, \hat{p}\left(p, b^j\right), y^{ij}, s, \varepsilon\right)$$

where *i* equals 1 if the price is known and *0* otherwise given the supply of price information b^j and conditional income is defined in terms of the cost of price information as $y^{ij} = y - c(i, b^j)$. The consumer will choose to know the price if

(4-6) $$D^{j*} = v^{1j} - v^{0j} \geq 0.$$

where D^* is defined as a latent variable that indexes the net utility of knowing the price given the available price information. Table 11 shows the four possible indirect utility outcomes for a discrete change in the supply of price information from b^1 to b^0. The analysis in this chapter focuses on the special case where b^0 is the level of price information that ensures everyone will know the price. In this case, cell (2, 2) in Table 11 is irrelevant because everyone will know the price after the change in price information.

The compensating variation (CV) of a discrete change in the supply of price information from b^1 to the level b^0 that generates accurate price perceptions is[6]

[6] In general, the CV of a discrete change in the supply of price information from b^1 to b^0 is given by

$$CV\left(b^1, b^0\right) = e\left(\hat{p}\left(p, b^1\right), v^{i1}, s, \varepsilon\right) - e\left(\hat{p}\left(p, b^0\right), v^{i1}, s, \varepsilon\right).$$

This formulation is general because $\hat{p}\left(p, b^0\right)$ does not necessarily equal the actual price. Thus, an individual may chose to know the price in either state of the world and can switch from not knowing to knowing or vice versa following the change in the price information.

$$(4\text{-}7)\quad \begin{aligned} CV\left(b^1,b^0\right) &= e\left(\hat{p}\left(p,b^1\right),v^{11},s,\varepsilon\right)-e\left(p,v^{11},s,\varepsilon\right) \\ &= pQ\left(\hat{p}\left(p,b^1\right),v^{11},s,\varepsilon\right)-pQ\left(p,v^{11},s,\varepsilon\right) \end{aligned}$$

where perceived price equals the actual price with the information level b^0 so that $p=\hat{p}\left(p,b^0\right)$. In this case of *perfect price information* the individual may or may not know the price before the change in price information, but after the change they *will* know the price. The CV measure can be recovered from observed expenditure patterns before and after the change. When there is no data available for behavior after the change in price information these outcomes must be simulated where needed. In this case, observations related to cell (2, 1) in Table 11 are not available so this information must be inferred from the data on behavior with the reference supply of price information (row 1). Chapter 2 reviews approaches to recovering welfare measures in this case. Two approaches are described and applied following the discussion of the value of price information.

Table 11. Utility outcomes with price knowledge and information change

		CHOOSE TO KNOW PRICE?	
		YES (i=1)	NO (i=0)
CHANGE IN PRICE INFORMATION	BEFORE (j=1)	$v\left(1,\hat{p}\left(p,b^1\right),y^{11},s,\varepsilon\right)$	$v\left(0,\hat{p}\left(p,b^1\right),y^{01},s,\varepsilon\right)$
	AFTER (j=0)	$v\left(1,\hat{p}\left(p,b^0\right),y^{10},s,\varepsilon\right)$	$v\left(0,\hat{p}\left(p,b^0\right),y^{00},s,\varepsilon\right)$

The value of price information is shown in Figure 4 and Figure 5 for the second and third cases of price perception, respectively, for the assumed form of $\hat{p}\left(p,\hat{b}\right)=p\cdot\hat{b}$. There are two compensated demands shown in each figure indicating that the slope and position of a demand curve depend on the available price information. Specifically, differentiating the compensated demand function with respect to the actual price p writes an expression for the slope of the demand curve in actual price/quantity space

(4-8)
$$\begin{aligned}\frac{\partial Q^c}{\partial p} &= \frac{\partial Q^c}{\partial \hat{p}}\frac{\partial \hat{p}}{\partial p} \\ &= \frac{\partial Q^c}{\partial \hat{p}}\hat{b} \qquad \textit{for } \hat{p}\left(p,\hat{b}\right) = p \cdot \hat{b}\end{aligned}$$

Thus, the slope of the perceived price demand curve will differ from the slope of the actual demand curve by a factor relating the perceived price to the actual price. Those who know the price can respond differently to the same change in actual price than those without price information. Note that total change in uncompensated demand with respect to a change in perceived price is manifest in the substitution effect

(4-9)
$$\begin{aligned}\frac{\partial Q^m}{\partial \hat{p}} &= \frac{\partial Q^c}{\partial p}\left(\frac{\partial \hat{p}}{\partial p}\right)^{-1} - \frac{\partial Q^m}{\partial y}Q^m \\ &= \frac{\partial Q^c}{\partial p}\hat{b}^{-1} - \frac{\partial Q^m}{\partial y}Q^m \qquad \textit{for } \hat{p}\left(p,\hat{b}\right) = p \cdot \hat{b}\end{aligned}$$

where Q^m is the uncompensated demand for Q.[7]

For the first case noted above ($\hat{b}$ = *1*), the slopes of the perceived and actual price demand curves are the same. Figure 4 illustrates the second case ($\hat{b} > 1$), where the consumer *overestimates* the actual price and the slope of the perceived price demand curve is *greater* than the slope of the actual price demand curve. Area A measures the value of price information (i.e., of a change in price perception) in (4-7). If more

---

[7] The version of the Slutsky equation in expression (4-9) is derived as shown in Varian (1992 p. 120), starting from compensated demand function with the perceived price argument. Only own price effects are illustrated, but these effects represent changes in relative value since, from above, the own price and income terms are normalized on a numeraire good.

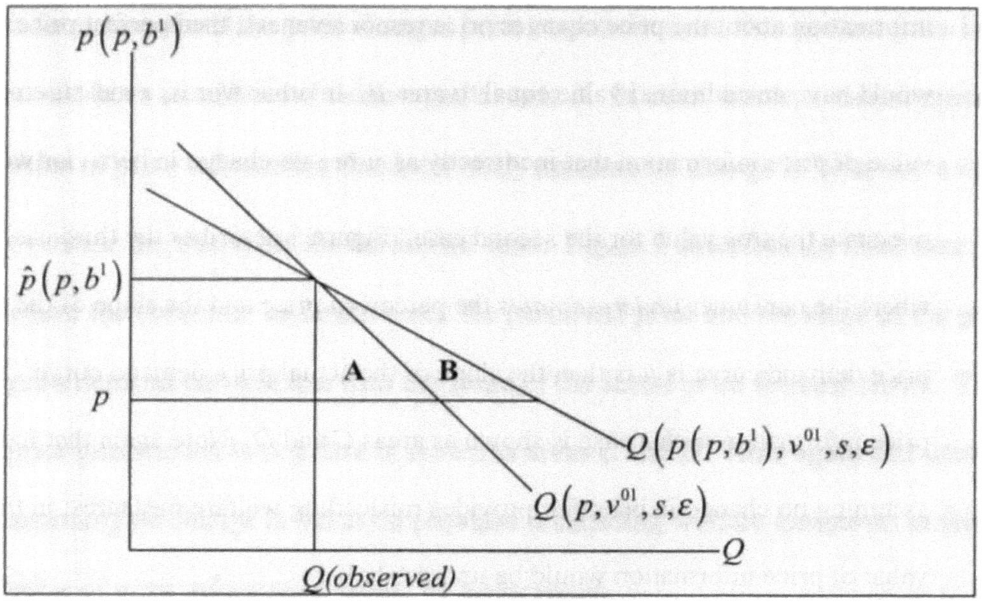

Figure 4. Value of price information: perceived price greater than actual price

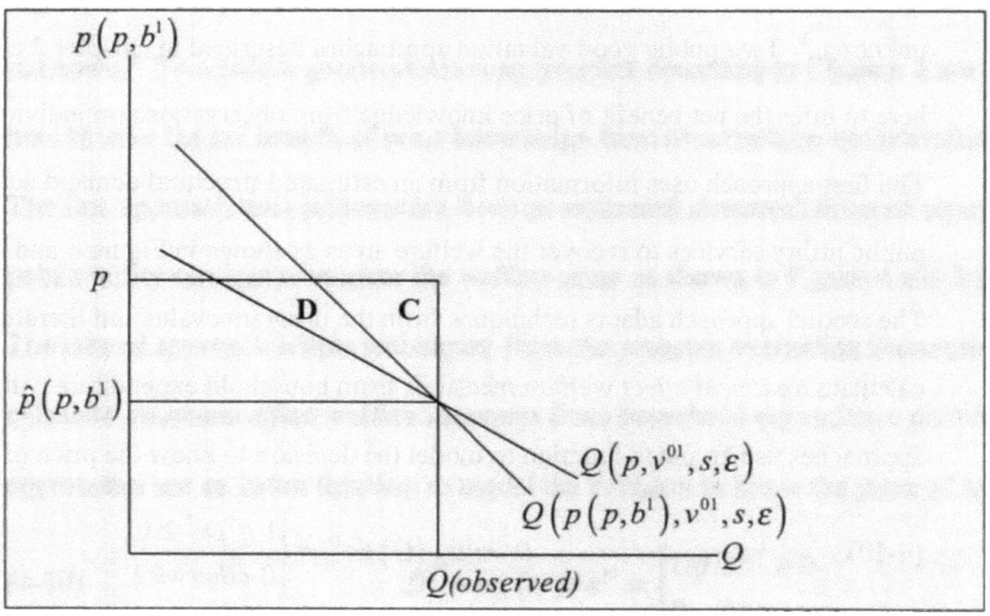

Figure 5. Value of price information: perceived price less than actual price

information about the price changes price responsiveness, then perfect price knowledge would have an additional value equal to area *B*. In other words, a measurement of the value of price information that incorrectly assumes no change in behavior would overstate the true value for the second case. Figure 5 describes the third case ($0 < \hat{b} < 1$), where the consumer *underestimates* the perceived price and the slope of the perceived price demand curve is *less* than the slope of the actual price demand curve. The value of price information in this case is shown as areas *C* and *D*. Note again that incorrectly assuming no change in behavior provides misleading welfare measures; in this case the value of price information would be understated.

Empirical Models

Households will chose to know the true price if the net benefit of doing so is non-negative, but the point at which an individual household will choose to do so is generally unknown.[8] Two public good valuation approaches described in Chapter 2 are applied here to infer the net benefit of price knowledge from observations on individual behavior. The first approach uses information from an estimated structural demand equation for public utility services to recover the welfare areas as shown in Figure 4 and Figure 5. The second approach adapts techniques from the program evaluation literature to calculate *treatment effect* welfare measures from household expenditure patterns. Both approaches use an index function to model the decision to know the price of service

(4-10)
$$D^j = \mu_D(G) + \varepsilon^D = \begin{cases} 1 \ if \ D^{j*} \geq 0 \\ 0 \ otherwise \end{cases}$$

[8] The empirical section of the chapter refers to the decision unit as a 'household' rather than a 'consumer' to be consistent with the case study data.

where $\mu_D(\cdot)$ is a function of a vector G of observed random variables that affect the decision to know the price and ε^D is an unobserved random variable that represents the relevant portions of ε. Note that D^{j*} is the latent net utility variable in (4-6) so this index function summarizes an individual's preferences over the decision to know the price or not.

Water Demand Model

Following Hanemann (1984a) the observed quantity of public utility services demanded before the price information change can be represented as

(4-11)

$$Q\left(\hat{p}\left(p,b^1\right),y,s,\varepsilon\right)=\begin{cases}\partial v^{11}/\partial\hat{p}\left(p,b^1\right) & \textit{if } D^1=1\\ \partial v^{01}/\partial\hat{p}\left(p,b^1\right) & \textit{otherwise}\end{cases}$$
$$=\begin{cases}Q\left(\hat{p}\left(p,b^1\right),y^{11},s,\varepsilon\right) & \textit{if } D^1=1\\ Q\left(\hat{p}\left(p,b^1\right),y^{01},s,\varepsilon\right) & \textit{otherwise}\end{cases}$$

With consistent specifications for the demand equation $Q(\ )$ and the observable portion of the index equation $\mu_D(\cdot)$, this model can be estimated as a switching regression before the change in price information (Maddala 1983).[9] The results can then be use to simulate the welfare measures defined in Figure 4 or Figure 5. However, another well-known estimation issue arises with public utility services where quantity and marginal price are determined simultaneously with block rate pricing structures.

Ideally, estimation in this context should proceed with techniques that also consider the simultaneity of the discrete block choice and continuous quantity choice decisions

---

[9] If the decision to know the price is not correlated with consumption decisions then OLS can be used to estimate separate demand equations for those who *know* and those who *don't know*. Otherwise, a technique to correct for endogenous sample selection must be used to get unbiased estimates of the parameters in the two demand equations. I hypothesize that the two decisions are, in fact, correlated and propose a general model of endogenous switching to estimate the demand equations for the two regimes.

(Cavanagh, Hanemann and Stavins 2001; Hewitt and Hanemann 1995; Reiss 2001; Rietveld, Rouwendal and Zwart 1997; Terza and Welch 1982).[10] I opt for a simpler approach, following Agthe et al. (1986) and Chicoine, Deller and Ramamurthy (1986), in specifying price and quantity equations as a system of simultaneous linear equations.[11] There is a cost associated with linearizing the budget constraint in this manner (Maddock, Castano and Vella 1992), but I choose this approach to keep the analysis manageable and to maintain the focus on the self-selection related to price knowledge.

Recognizing that the actual price and quantity are jointly determined with block rate pricing, the model in (4-10) and (4-11) can be respecified as a simultaneous equations model with endogenous selectivity (Lee, Maddala and Trost 1980):

$$D^1 = G\delta + \varepsilon^D \tag{4-12}$$

$$\left.\begin{aligned} P^{11} &= \omega^{11}H + \varphi^{11}\lambda^{11} + \upsilon^{11} \\ Q^{11} &= Q\left(\hat{P}\left(\tilde{P}^{11}, b^1\right), y^{11}, s\right) + \rho^{11}\lambda^{11} + \varepsilon^{11} \end{aligned}\right\} \quad \textit{if} \quad D^1 = 1 \tag{4-13}$$

$$\left.\begin{aligned} P^{01} &= \omega^{01}H + \varphi^{01}\lambda^{01} + \upsilon^{01} \\ Q^{01} &= Q\left(\hat{P}\left(\tilde{P}^{01}, b^1\right), y^{01}, s\right) + \rho^{01}\lambda^{01} + \varepsilon^{01} \end{aligned}\right\} \quad \textit{if} \quad D^1 = 0 \tag{4-14}$$

where v^{jl} and ε^{jl} are the jointly normal error terms in the price and demand regressions, respectively. The model consists of five estimating equations.[12] First, the binary indicator variable D^l is used as the dependent variable in a probit regression on G to yield

---

[10] Although one might question the applicability of estimation frameworks that precisely model the choices with nonlinear budget constraints when consumers are not fully aware of the rate schedule and their use levels. These approaches assume that consumers have enough information to simultaneously choose or act *as-if* they are choosing the efficient block to consume in and the level to consume within the block.

[11] The studies of water demand using instrumental variable techniques also follow this general approach (Deller, Chicoine and Ramamurthy 1986; Jones and Morris 1984).

[12] The general form of $Q(\ )$ is maintained to simplify the notation. The exact form of the demand equations to be estimated is presented below.

the estimated parameter vector δ. Following the standard procedure for two-stage sample selection models, the predicted values of the probit equation are used to compute

$$\lambda^{11} = \phi\left(G\hat{\delta}\right)/\Phi\left(G\hat{\delta}\right) \text{ and } \lambda^{01} = -\phi\left(G\hat{\delta}\right)/\left[1-\Phi\left(G\hat{\delta}\right)\right]$$

for those who know ($D^I = 1$) the price and those who don't know ($D^I = 0$) the price, respectively. The inverse Mills ratio variables λ^{i1} are included as regressors along with the vector H of explanatory variables in the two price equations. The regression produces the estimated price equation coefficients ω^{iI} and φ^{iI} for each group. Note that the φ^{iI} parameter is a measure of covariance between the price information decision equation and the price equation. A test of statistical significance for φ^{ij} indicates whether the price information decision is endogenous with respect to the (linearized) price choice decision. The choice of the variables H for the price equations is somewhat arbitrary since these expressions serve as crude approximations of the true discrete relationship between price and quantity. I use the specification of Terza (1986) and Nieswiadomy and Molina (1989) that forms H with the exogenous variables in $Q(\ )$ and the prices that a given household would face at four different levels of monthly (water) consumption (6000, 12000, 18000, and 24000 kgals). Finally, the predicted prices $\tilde{P}^{i1}$ are used along with λ^{iI} in estimation of the two demand equations to produce the estimated coefficients and ρ^{iI} for each group. Like φ^{iI}, the ρ^{iI} parameter is a measure of covariance between the price information decision equation and the demand equation. A test of statistical significance for ρ^{iI} indicates whether the price knowledge decision is endogenous with respect to the demand decision. The estimation of equations (4-12) through (4-14) will produce consistent estimates of the parameters in the demand equations and can be executed easily with a procedure shown in Greene (1995 pp. 643-44).

The empirical specification of the $Q(\ )$ and $\hat{P}$ functions used were developed by Shin (1985) to measure price perception among utility service customers. The perceived price $\hat{p}$ is defined as a function of lagged average price, AP_{t-1}, marginal price, MP, and a price perception parameter k such that

$$\hat{P} = MP\left(\frac{AP_{t-1}}{MP}\right)^k \tag{4-15}$$

which, implies that the stock of exogenously supplied price information is a function of the actual price $p = MP$ and the previous period's average price of service.[13] As constructed, the value of the perceived price variable depends on the parameter k: if the consumer only responds to marginal price, then $k = 0$, and if they only respond to lagged average price, then $k = 1$. Values for k between zero and one imply that the perceived price is between marginal and average price, while values outside of this range suggest that the consumer is responding to some other price level.

In double-log form, the partial adjustment[14] model estimated by Shin (1985) and Nieswiadomy and Molina (1991) appears as:

$$\begin{aligned} Ln(Q) &= \alpha + Ln(Q_{t-1})(1-\theta) + Ln\left[MP(AP_{t-1}/MP)^k\right]\theta\eta \\ &\quad + Ln(y)\theta\nu + Ln(s)\theta\beta \\ &= \alpha + Ln(Q_{t-1})(1-\theta) + Ln(MP)\theta\eta + Ln\left[(AP_{t-1}/MP)\right]k\theta\eta \\ &\quad + Ln(y)\theta\nu + Ln(s)\theta\beta \end{aligned} \tag{4-16}$$

---

[13] Note that all prices and income levels are adjusted to *relative* values by dividing the monthly income and price variables by a regional CPI for the month of observation. This adjustment is necessary to preserve the homogeneity restriction and ensure that the estimated demand equation is consistent with utility maximization (Hanemann 1998).

[14] The partial adjustment model is used because households are unable to fully adjust their water use in the short run (billing cycle) given a fixed stock of water-using capital. I explicitly write out the model with the partial adjustment parameter as it appeared in Houthakker, Verleger and Sheehan (1974). Note that all terms without '*t-1*' subscripts indicate values at time *t*.

where α is the intercept, θ is the partial adjustment parameter (to be estimated), η is the constant perceived price elasticity, γ is the constant income elasticity, and β is a conformable vector of parameters on s. I suggest an alternative formulation of the same model that separates the $Ln(MP)$ and $Ln(AP_{t-1})$ elements of the $Ln(AP_{t-1}/MP)$ term:

$$\begin{aligned} LnQ &= \alpha + Ln(Q_{t-1})(1-\theta) + Ln(MP)\theta\eta(1-k) + Ln(AP_{t-1})k\theta\eta \\ &\quad + Ln(y)\theta\nu + Ln(s)\theta\beta \end{aligned} \tag{4-17}$$

This form of the model reveals that Shin's perceived price signal is simply a weighted combination of the marginal price and lagged average price signals. Using the fact that $AP_{t-1} = BILL_{t-1}/Q_{t-1}$, the model can be simplified further to yield[15]

$$\begin{aligned} LnQ &= \alpha + Ln(Q_{t-1})\left[(1-\theta)(1+k\eta)\right] + Ln(MP)\theta\eta(1-k) + Ln(BILL_{t-1})k\theta\eta \\ &\quad + Ln(y)\theta\nu + Ln(s)\theta\beta \end{aligned} \tag{4-18}$$

This is essentially a short run specification, but the long run effects can be recovered by manipulating the partial adjustment parameter. Full adjustment ($\theta = 1$) occurs subject to the household's price perception. If the household maintains a perceived price other than their marginal price ($k \neq 0$), then they cannot adjust Q to an efficient level relative to the consumption of other items. The completely adjusted household ($\theta = 1$, $k = 0$) will have optimally selected its capital stock and be reacting to changes in the marginal price of service. For reference the formulas used to recover the key parameters of interest are shown in Table 12. The derivation of the value of price information measures from the demand model parameters is described in the Results.

---

[15] $BILL_{t-1}$ is a nonlinear function of Q_{t-1} because of the block nature of the rate structures so the there is less chance of introducing multicollinearity by including both variables as regressors.

Table 12. Formulas for key model parameters

Ln(Q) = [b0] + [b1]Ln(Q^{t-1}) + [b2]Ln(MP) + [b3]Ln($BILL^{t-1}$) + [b4]Ln(Y) + [β]Ln(X)

Parameter	Short-Run	Long-Run
Income elasticity	b4	b4(1-b1)
Perceived price elasticity	b2+b3	(b2+b3)/(1-b1)
Marginal price elasticity	b2	b2/(1-b1)
Price perception (k)	b3/(b2+b3)	b3/(b2+b3)
Partial adjustment (θ)	1 – (b1/(1+ b3))	1

Treatment Effects Bill Model

The general *treatment effects* framework introduced in Chapter 2 is an alternative way to recover the value of price information. This framework is designed to evaluate welfare measures for public good changes with expenditure data for different segments of population. For the present case, I am suggesting that there is a hypothetical program to *fully* inform consumers about the price of Q. The additional price information can be considered a change in the supply of a public good from b^1 to b^0 such that $\hat{p}(p,b^0) = p$. Formally, the difference in an individual's monthly bill with and without the additional price information is the *treatment effect* of price information program

$$\begin{aligned} \Delta &= e\left(\hat{p}\left(p,b^1\right),v^{i1},s,\varepsilon\right) - e\left(\hat{p}\left(p,b^0\right),v^{i0},s,\varepsilon\right) \\ &= e\left(\hat{p}\left(p,b^1\right),y,s,\varepsilon\right) - e\left(\hat{p}\left(p,b^0\right),y,s,\varepsilon\right) \\ &= BILL^{i1} - BILL^{i0} \end{aligned} \quad (4\text{-}19)$$

where the first line suggests that that this spending difference is an uncompensated measure of the value of price information shown in (4-7). This measure can be directly recovered with observations on the two spending outcomes for each individual. The approach developed in Chapter 2 can recover the welfare measures in (4-19) and (4-7) when observations on spending outcomes are only available for a period before the change in price information. The approach applies to the case where every individual knows the price with the program information b^0. For this special case, the welfare measures can be recovered from the difference in spending with and without price knowledge

$$\begin{aligned} \hat{\Delta} &= e\left(\hat{p}\left(p,b^1\right),v^{01},s,\varepsilon\right)-e\left(\hat{p}\left(p,b^1\right),v^{11},s,\varepsilon\right) \\ &= e\left(\hat{p}\left(p,b^1\right),y,s,\varepsilon\right)-e\left(p,y,s,\varepsilon\right) \\ &= BILL^{01} - BILL^{11} \end{aligned} \tag{4-20}$$

The *treatment effects* approach designates those in the sample who *choose* to know the price as the *treatment group* and those who do not the *control group*. Since individuals *self-select* into one group or the other, the difference in expenditure patterns among groups is important.

There are three possible sources of differences in expenditures between the price information treatment and control groups: 1) price knowledge may lead to higher (or lower) average expenditure on the commodity, 2) those who know the price may spend more (or less) on the commodity in the first place, and 3) the expenditure of those who know the price may have changed more (or less) because of the price knowledge than those who don't know the price, if they did. As described in Chapter 2, the information inherent in these sources of variation can be used to infer the relative value of the decision to learn the price (Heckman 2001a). Specifically, the decision to know the price or not modeled by the index (4-10) can be combined with the expenditure information to recover the welfare measures in (4-7) and (4-19).

In the treatment effects approach (TEA), expenditures are modeled as an endogenous switching regression

$$\text{(4-21)} \quad BILL = \begin{cases} BILL^{11} & if \quad D^1 = 1 \\ BILL^{01} & otherwise \end{cases}$$

where $j = 1$ because b^1 is used as the reference state of price information and D^1 is the price knowledge index equation defined in (4-10). Expenditure for the monthly public utility services bill is modeled with a linear Engel relationship:

$$\text{(4-22)} \qquad BILL^{i1} = \beta_I^{i1} + BILL_{t-1}\beta_{BILL_{t-1}}^{11} + y\beta_y^{i1} + s\beta_s^{i1} + \varepsilon^{i1}.$$

where, following Phlips (1983), income is normalized to the own price level and the intercept and error term for the public good users ($i = 1$) are implicitly defined as $\beta_I^{11} = \tilde{\beta}_I^{11} p^{11} + \tilde{\beta}_p^{11} p^{01}$ and $\varepsilon^{11}p^{11}$, respectively. These terms are defined similarly for the nonusers ($i = 0$). Thus, prices appear endogenously as a portion of the unobservable determinants of spending. The lagged dependent variable appears as a regressor because, as was assumed in the structural demand model, a household's water use and spending can only partially adjust towards their optimal consumption level in the short run.

Following the procedure in Chapter 3, (4-22) is integrated to recover the related indirect utility function necessary to specify the form of the related net utility index D^1. In the Appendix I show that the resulting net utility index can be reduced to a simple linear in variables equation with an additive error. The full endogenous switching estimating system is

$$\text{(4-23)} \qquad D^1 = G\beta^{D^1} + \varepsilon^{D^1}$$

$$\text{(4-24)} \qquad BILL^{11} = X\beta^{11} + \varepsilon^{11}$$

$$\text{(4-25)} \qquad BILL^{01} = X\beta^{01} + \varepsilon^{01}$$

where the constant and socioeconomic variables are collected in $X = (1, BILL_{t-1}, y, s)$ and $G = (1, y, s, z)$, and the conformable parameter vectors are $\beta^{i1} = (\beta_I^{i1}, \beta_{BILL_{t-1}}^{i1}, \beta_y^{i1}, \beta_s^{i1})$ for the spending equations and $\beta^{D^i} = (\beta_I^{D^i}, \beta_y^{D^i}, \beta_s^{D^i}, \beta_z^{D^i})$ for the index equation. Note that *z* represents the exclusion restriction required for the general index equation specification (Heckman and Vytlacil 2001a). Assuming ε^{D^i}, ε^{11}, and ε^{01} are joint normally distributed, the FIML estimates the of the parameters $\{\beta^{11}, \beta^{01}, \beta^{D^i}, \sigma^{11}, \sigma^{01}, \rho^{11D^i}, \rho^{01D^i}\}$ are obtained using the endogenous switching estimator in LIMDEP (Greene 1995).[16] The parameter definitions are described in the results.

As defined in Chapter 2 and implemented in Chapter 3, the standard treatment effect and the policy relevant treatment effect measures for this model are (Heckman, Tobias and Vytlacil 2001):

(4-26) $$ATE^{\hat{\Delta}} = E\left[\hat{\Delta} | y, s\right] = X\left(\beta^{11} - \beta^{01}\right)$$

(4-27) $$TT^{\hat{\Delta}} = E\left[\hat{\Delta} | y, s, D = 1\right] = X\left(\beta^{11} - \beta^{01}\right) + \left(\sigma^{11D} - \sigma^{01D}\right)\lambda^{11}$$

(4-28) $$TU^{\hat{\Delta}} = E\left[\hat{\Delta} | y, s, D = 0\right] = X\left(\beta^{11} - \beta^{01}\right) + \left(\sigma^{11D} - \sigma^{01D}\right)\lambda^{01}$$

(4-29) $$E\left[\hat{\Delta}^{*}\right] = X\left(\beta^{11} - \beta^{01}\right) + \left(\sigma^{11D} - \sigma^{01D}\right)\left[-G\beta^{D}\right]$$

where $\lambda^{11} = \phi\left(G\hat{\beta}^{D}\right) / \Phi\left(G\hat{\beta}^{D}\right)$ and $\lambda^{01} = -\phi\left(G\hat{\beta}^{D}\right) / \left[1 - \Phi\left(G\hat{\beta}^{D}\right)\right]$ are the inverse Mills ratios for the spending equations with and without price knowledge, respectively. Expression (4-29) is the policy relevant treatment effect welfare measure for *CV*. This

---

[16] The covariances are easily recovered from the correlation coefficient because the variance of the index equation is normalized to unity.

calculation evaluates the difference in spending at the point where, based on the index equation, the individual is just indifferent to knowing the price or not. See Chapter 2 for a full discussion of this welfare measure.

Heckman, Tobias and Vytlacil (2001) show simple unconditional estimators for each of the four treatment effect parameters as

$$E[K] \approx \frac{1}{N}\sum_{n}^{N} K(X_n, z_n) \tag{4-30}$$

where K is the treatment effect measure of interest and N is the number of observations in the relevant set, i.e., N is the whole sample for (4-26) and (4-29), only the users for (4-27) and only the nonusers for (4-28). Note that expectation in (4-29) can be conditioned on any subset of the sample. For example, evaluating (4-29) over those who know the price, gives the expected treatment effect welfare measure for a randomly chosen individual from this group. This calculation and a similar one for those who don't know the price is reported in the results.

Data

The data set is composed of responses from a 1997 survey of North-Central Florida households (BEBR 1997) and the corresponding 1997-99 monthly billing records from the water/sewer utilities in Gainesville, Ormand Beach, and Cocoa Beach. Monthly precipitation and temperature records from the nearest regional airport for each utility were also added to each observation. Missing data for all variables, except water use, were set to the mean values observed in the relevant utility service area. All monthly observations with no reported water use for two consecutive months in Gainesville and

Cocoa Beach were dropped from the sample.[17] This was done primarily to accommodate the double-log demand model specification, but it can be rationalized to the extent that zero water usage is caused by factors exogenous to water use decisions (e.g., vacations). Ormand Beach has a fixed monthly allowance of two thousand gallons so, following the logic of Hewitt (2000), I set any observed water use below this amount to the allowance and set the marginal price to the uniform charge. The final sample consisted of 742 individual households divided about evenly among the three utility areas.

Table 13 presents the rate schedules for the three utilities in the sample. Note that only one absolute change in charges over the sample period occurred (in Ormand Beach) so that the price variation in the sample is mainly cross-sectional in nature. The marginal price at Ormand Beach is a simple uniform charge, but the rate schedules at Gainesville and Cocoa Beach are somewhat peculiar because of caps on the amount of monthly water use that can be billed at the sewer rate.[18] Oddly, the sewer cap in Cocoa Beach causes the combined water and sewer marginal price to increase up to the cap, decline immediately after the cap, and then increase again. The rate schedule at Gainesville has similar complications due to seasonal charges and a sewer cap that varies per household based on their maximum *winter* usage. The differences in rate schedules suggest that it may cost less to determine the marginal price at Ormand Beach than for the other two utilities.

---

[17] As a consequence of dropping the zero water use observations, some households will not have observations for every month in the study period. All required data transformations (variable lags, etc.) were made prior to eliminating any observations in order to maintain the integrity of the data.

[18] The prices used in the study include charges for both water and sewer service, because the monthly bills presented to the water users reflect the joint costs of these services. Griffin and Chang (1990) introduced the potential estimation issues with combined water and sewer rate structures, but there has been little, if any, published discussion since.

Table 13. Rate schedules in study area

Blocks		$/1000 gallons[a]			$/month		
		Water	Sewer	Total[b]	Water	Sewer	Total[b]
Ormand Beach							
Before 10/97	0 - 2	0.00	0.00	0.00	7.68	10.02	17.70
	> 2	1.93	2.70	4.63			
After 10/97	0 - 2	0.00	0.00	0.00	8.06	10.52	18.58
	> 2	2.03	2.84	4.87			
Cocoa Beach							
0 - 8		1.36	2.80	4.16	7.99	6.00	13.99
9 - 12		1.58	2.80	4.38			
13 - 16		1.58	0.00	1.58			
17 - 24		1.90	0.00	1.90			
> 24		2.56	0.00	2.56			
Gainesville[c]							
November - March	All Usage	0.98	2.43	3.41	3.00	2.11	5.11
April - October	0 - 9	0.98	2.43	3.41			
	> 9	1.29	2.43	3.72			

[a]All charges are for water and sewer service and are shown as nominal values. They were adjusted to relative monthly values before estimation with the monthly all item CPI for Southern cities of size class B/C.
[b]Some households in the sample are not connected to the sewer system. The total marginal and monthly charges for these observations only reflect the cost of water service.
[c]The complete rate structure for Gainesville is not shown in the table because sewer charges are conditional on a household's *maximum winter usage*. This household specific quantity is determined in the months of January and February and forms a cap on billable sewage for the rest of the year.

Table 14 lists the summary statistics over all households and months for the other relevant study data according to a household's reported price knowledge. Only 6% of the households in the sample reported that they knew the marginal price for water service. Interestingly, these households used about a thousand gallons less water on average each month than those who did not know the marginal price. The former also faced a lower marginal price on average than the latter, which may be due to increasing block features

Table 14. Summary statistics for water demand data

Variable	Name	Know Price	Don't Know Price	all
Monthly Consumption (1000 Gallons)[a]	KGAL	8.310	9.340	9.270
		(8.540)	(9.840)	(9.760)
Bill ($/Month)[a, c]	BILL	28.120	35.450	34.980
		(21.380)	(30.850)	(30.380)
Monthly Income ($)[b, c]	INCOME	4,567	4,460	4,467
		(2,846)	(3,344)	(3,314)
Lawn Size (Acres)[b]	LAWN SIZE	0.530	0.490	0.490
		(0.540)	(0.600)	(0.600)
Household Size[b]	HOUSEHOLD SIZE	2.920	3.170	3.160
		(0.940)	(1.180)	(1.170)
Monthly Mean Temp (Degrees F)[d]	TEMPERATURE	69.610	69.590	69.590
		(10.340)	(10.340)	(10.340)
Monthly Precipitation (Inches)[d]	PRECIPITATION	4.350	4.360	4.360
		(3.220)	(3.210)	(3.210)
Drink Bottled Water?[b]	BOTTLED WATER	0.450	0.390	0.390
		(0.500)	(0.490)	(0.490)
Heard About Low Flow Fixtures?[b]	LOW FLOW	0.770	0.620	0.630
		(0.420)	(0.480)	(0.480)
Heard About Xeriscaping?[b]	XERISCAPE	0.430	0.360	0.370
		(0.490)	(0.480)	(0.480)
Ormand Beach Resident?	ORMAND BEACH	0.380	0.320	0.330
		(0.480)	(0.470)	(0.470)

Standard deviations in parentheses.
[a]From utility billing records.
[b]Reported in the water customer survey (BEBR 1997).
[c]Adjusted by the U.S. Bureau of Labor estimates of the 1997-99 monthly CPI (1996 = 100) for all items in southern U.S. cities with populations less than 1.5 million (size B/C).
[d]Measurements from the nearest regional airport stations as reported by the U.S National Oceanic & Atmospheric Administration.

in two of the sample rate structures.[19] I assume that all variables, except water use, price, bill, temperature, and precipitation, are fixed for the study period. That is, the household values for these variables are assumed to be the same in 1998 and 1999 as they were reported in the 1997 survey. This assumption is plausible for the basic socioeconomic variables, such as household size and lawn size, but is somewhat tenuous for the variables relating to bottled water consumption and knowledge of conservation practices. For example, a household that did not consume bottled water in 1997 may have started to consume this product in 1998 or 1999. More critically for present purposes, though, is the possibility that a household who did not know the marginal price of water service in 1997 may have actually learned the price in the subsequent years. In this case the results will be conservative approximations of the actual statistical differences in water use behavior between the *know* and *don't know* groups.

Results

Structural Demand Model

The estimation results for the price information decision equation (Info Decision) and the price/demand equations for each knowledge regime are shown in Table 15. In the probit estimation (far left column) of the price information decision equation, all variables, except income, are significant at the 5 percent level. Income is probably not a statistically important factor in the probability of knowing the marginal price because water bills constitute a small share of household income. The probability of knowing the marginal price increases with lawn size, but decreases with household size. It may be that people with larger lawns have a greater interest in the cost of irrigating and thus, the

[19] Simple t-tests indicate that mean water use and price, as well as the means of several socioeconomic variables are significantly different between the two the *know* and *don't know* samples.

Table 15. Water demand model estimation results

Variable	Info Decision	Know Price		Don't Know Price	
		Price	Demand	Price	Demand
Constant	-1.45E+00 (1.56E-01)*	-7.92E-01 (5.06E-01)	-1.06E+00 (6.62E-01)	-3.28E-01 (1.14E-01)*	-8.65E-01 (1.00E-01)*
INCOME	1.01E-02 (1.81E-02)	-2.72E-02 (2.65E-02)	7.61E-02 (3.16E-02)*	-1.16E-02 (4.93E-03)*	7.74E-02 (4.62E-03)*
LAWN SIZE	1.57E-01 (2.13E-02)*	1.58E-02 (2.53E-02)	-1.55E-02 (3.99E-02)	5.54E-03 (6.89E-03)	1.96E-02 (6.60E-03)*
HOUSEHOLD SIZE	-3.16E-01 (3.87E-02)*	1.64E-01 (5.83E-02)*	2.91E-01 (8.00E-02)*	1.21E-02 (1.26E-02)	1.06E-01 (1.17E-02)*
BOTTLED WATER	1.48E-01 (2.56E-02)*				
LOW FLOW	2.96E-01 (2.97E-02)*				
XERISCAPE	6.33E-02 (2.69E-02)*				
ORMAND BEACH	1.08E-01 (2.64E-02)*				
$BILL_{t-1}$		6.68E-01 (6.41E-02)*	1.98E-01 (1.25E-01)	5.79E-01 (1.82E-02)*	1.42E-01 (2.04E-02)*
TEMPERATURE		7.89E-02 (9.22E-02)	2.37E-01 (1.22E-01)*	-3.24E-02 (2.36E-02)	1.08E-01 (2.03E-02)*
PRECIPITATION		-6.85E-04 (1.17E-02)	-4.70E-02 (1.77E-02)*	1.65E-02 (3.08E-03)*	-3.99E-02 (2.95E-03)*
Q_{t-1}		-4.48E-01 (3.74E-02)*	5.99E-01 (8.74E-02)*	-4.77E-01 (1.04E-02)*	6.81E-01 (1.51E-02)*
MPT @ 6,000 gal.		2.03E-01 (4.02E-02)*		1.26E-01 (1.02E-02)*	
MPT @ 12,000 gal.		-7.74E-02 (4.94E-02)		1.13E-01 (1.20E-02)*	
MPT @ 18,000 gal.		-2.42E-01 (1.09E-01)*		-1.39E-01 (1.95E-02)*	
MPT @ 25,000 gal.		2.69E-01 (7.44E-02)*		2.45E-01 (1.49E-02)*	
MPT			-2.83E-01 (1.02E-01)*		-1.76E-01 (1.69E-02)*
λ^{ii}		-1.44E-01 (1.07E-01)	-2.69E-01 (1.48E-01)*	1.43E-02 (1.04E-01)	-1.13E-02 (9.15E-02)

All non-binary variables are in logarithms.
Standard errors are shown in the parentheses below each estimate.
*Parameter estimate significant at the 10 percent level.

price of water. For large households, knowledge of the marginal price may be less likely if water takes on public good characteristics such that one household member is unable to influence the water use of others. Knowledge of low flow fixtures and xeriscaping increases the probability of knowing the marginal price, as does the regular consumption of bottled water. These variables represent *water awareness* factors and the positive correlation with the price awareness variable establishes the consistency of household responses. A dummy variable is included for the utility (Ormand Beach) with the simplest rate structure. The parameter on this variable is positive indicating that (all else equal among utilities) households facing a simple rate structure are more likely to say they know the marginal price.

The results of the price instrument estimations are reported in the second and fourth columns of Table 15. These estimates are not the primary focus here, but note that the parameters φ^{11} and φ^{01} on the λ^{11} and λ^{01} terms, respectively, are not statistically significant in the price equations. This indicates that separate OLS estimation of the price equations with this dataset would not be subject to sample selection bias.

Of central interest in this chapter are the results from the demand equation estimations (columns 3 and 5 in Table 15). The parameter σ^{11} on the λ^{11} term in the 'know' price equation is significant at the 10 percent level and the σ^{01} parameter on the λ^{01} term in the 'don't know' price equation is not significant. Thus, the covariance between the decision equation and the demand equation is (weakly) significant for those who know the price, but not for those who do not. So the decision to learn the marginal price may be related to the water use decision if a household actually chooses to learn the price.

The parameter estimates for those who know the price are roughly comparable to the estimates for those who don't, but there are a few key differences. Most importantly, the parameters on the marginal price (MPT) and average price ($BILL_{t-1}$) terms are different between groups. The coefficient on the marginal price for the *know* group is over fifty percent larger than same coefficient for the *don't know* group. Furthermore, the coefficient on the average price term is not statistically significant for the *know* group, but is significant and positive for the *don't know* group. So the data for those who say they know the marginal price exhibits a statistical relationship between marginal price and water use, but not a relationship between average price and water use. We cannot necessarily say that the *know* group is actually *responding* to marginal price, but these results lend support to this possibility.

The results also suggest that the *know* group is more price sensitive than the *don't know* group and that the latter may actually respond to some combination of marginal and average price information. The short-run and long-run elasticities are shown in Table 16 for comparison. To explore the second premise, I used the formulas in Table 12 to calculate the price perception parameter and the perceived price at the mean values for the *don't know* group. The price perception parameter for the *don't know* group is $k = -4.10$, giving a perceived price of $0.33.[20] This perceived price is not statistically different from zero at a ten percent confidence level, suggesting that those who do not know the price of water service could be behaving as if water were free. Again, though, we cannot attribute causality to these statistical relationships. This interpretation should also be viewed cautiously as the estimated price perception parameter for the *don't know* group is

[20] The asymptotic standard errors of the price perception parameter and perceived price for the *don't know* group are, respectively, 1.49 and 0.267.

roughly four times larger than any value reported in previous studies (Shin 1985; Nieswiadomy 1992; Nieswiadomy and Cobb 1993; Nieswiadomy and Molina 1991).

Table 16. Estimates for key water demand model parameters

Parameter[a]	Short-Run		Long-Run	
	Know	Don't Know	Know	Don't Know
Income elasticity	0.076 (0.032)*	0.077 (0.005)*	0.190 (0.091)*	0.243 (0.018)*
Perceived price elasticity	-0.085 (0.056)	-0.035 (0.009)*	-0.213 (0.167)	-0.108 (0.031)*
Marginal price elasticity	-0.283 (0.102)*	-0.176 (0.017)*	-0.705 (0.135)*	-0.553 (0.031)*
Price perception	-2.312 (2.654)	-4.103 (1.499)*	-2.312 (2.654)	-4.103 (1.499)*
Partial adjustment	0.165 (0.054)*	0.173 (0.008)*	1.000 -	1.000 -

Standard errors are shown in the parentheses below each estimate.
[a]See Table 12 for corresponding formulas.
*Parameter estimate significant at the 5 percent level.

The econometric results indicate that those in the sample who do not know the marginal price for water are responding to a price lower than the actual price. This corresponds to the situation in Figure 5 where the value of price information is shown as areas *C* and *D*. The information from the demand estimations is used to calculate these areas as follows. The rectangle of total monthly (variable) expenditures on water/sewer service is calculated as $VE = (P_{actual} - P_{percieved})*Q_{observed}$. Then the estimated parameters are used to calculate the change in consumer surplus between the perceived and actual price for *two* demand situations using Hausman's (1980) exact measure. The first demand situation uses the information from the *don't know* estimation and assumes that the marginal price behavior (i.e., elasticity) would remain unchanged if they really knew the price. For the second situation I speculate that the price behavior of the *don't know* group would be more like the *know* group if they did actually know the price. To approximate this second case the perceived price elasticity from the *know* group is used in the consumer surplus calculations for the *don't know* group. Note that the constant term in the demand equation is normalized so that the demand lines intersect at ($Q_{observed}$, $P_{percieved}$) as shown in Figure 5. Subtracting the first demand consumer surplus measure from the estimated variable expenditures *VE* gives area *C* and subtracting the second consumer surplus measure from the first gives area *D*.

The aforementioned calculations are performed for short and long run effects. In the short run the over-expenditure due to lack of price knowledge (area *C*) was $0.94 per month for the average household that did not know the marginal price. If the price information actually caused the average household to change their price behavior, then the additional value of the price knowledge (area *D*) would be $1.29 per month. However, this area *D* calculation has a relatively large standard error ($0.842) and a Wald test indicates that it is not statistically significant from zero at the 10 percent level. The value of area *C* is only 0.02 percent of monthly income for the average household and the sum of areas *C* and *D* constitute about 0.05 percent, indicating that the value of price information is very small relative to income.

The long run over-expenditure due to lack of price knowledge was slightly higher at $0.97 per month for the average household, again about 0.02 percent of monthly income. However, the value of price information, assuming a change in price behavior, was considerably less in the long run at $0.90. This result appears because the partial adjustment parameter for those who know the price is smaller than the one for those who don't know the price. Intuitively, those households without price information are further off their optimal expenditures on water and can adjust relatively more in the long run. Compared to the short run value, the long run value of areas *C* and *D* is a slightly smaller portion, 0.04 percent, of monthly income for the average household.

Treatment Effects Model

The estimation results for the treatment effects model are shown in Table 17. The significant parameters in the price information decision equation are similar in sign and magnitude to those estimated in the water demand model. This is expected as the probit estimates shown in the water demand model (Table 15) are used as (part of) the starting values for the FIML estimation of the treatment effects model. Interpretation of these results follows the discussion presented for the water demand model.

The main interest in the treatment effects model is the estimates for the expenditure equations shown in the second two columns of Table 17. Although the magnitude of the estimates for these equations model are not directly comparable with the water demand model in Table 15, the signs are remarkably similar. The only anomaly among significant parameters is the coefficient on lawn size for the *don't know price* results. This parameter is positive in the water demand model (Table 15) and negative in the treatment effect bill model (Table 17). Thus, while those with larger lawns use more water, their bills tend to be less than those with relatively smaller lawns. This discrepancy likely occurs because the treatment effects model does not *control* for the variation in charges faced by households in the cross-section. The water demand model does control for differences in charges across service areas with prices among the independent variables. Since these variables are not explicitly included as regressors in the treatment effects model, the lower expenditures attributed to larger lawn sizes may actually occur because most of the households with larger lawns just happen to live in an

Table 17. Monthly bill treatment effects model results

Variable/Parameter	Info Decision	Know Price	Don't Know Price
CONSTANT	-1.57E+00	2.23E+00	1.61E+00
	(5.07E-02)*	(5.65E+00)	(1.10E+00)
INCOME	-4.21E-06	4.19E-04	3.86E-04
	(4.73E-06)	(1.71E-04)*	(2.00E-05)*
LAWN SIZE	8.05E-02	-8.77E-01	-9.90E-01
	(2.38E-02)*	(7.30E-01)	(2.60E-01)*
HOUSEHOLD SIZE	-9.82E-02	1.03E+00	5.11E-01
	(1.37E-02)*	(5.13E-01)*	(1.09E-01)*
BOTTLED WATER	1.53E-01		
	(2.64E-02)*		
LOW FLOW	2.96E-01		
	(3.16E-02)*		
XERISCAPE	6.25E-02		
	(2.85E-02)*		
ORMAND BEACH	1.00E-01		
	(2.71E-02)*		
$BILL_{t-1}$		7.35E-01	8.28E-01
		(8.90E-03)*	(6.20E-04)*
TEMPERATURE		9.08E-02	4.47E-02
		(3.84E-02)*	(1.27E-02)*
PRECIPITATION		-3.29E-01	-3.21E-01
		(1.10E-01)*	(3.86E-02)*
σ^{expend}		1.37E+01	1.67E+01
		(3.95E-01)*	(1.53E-02)*
$\rho^{selection,expend}$		-1.57E-01	1.79E-02
		(2.08E-01)	(1.90E-01)
$\sigma^{selection,expend}$		-2.15E+00	3.00E-01
		(2.91E+00)	(3.19E+00)

Standard errors are shown in the parentheses below each estimate.
*Estimate significant at the 0.10 level.
The final value of the log likelihood function is -110038.4.

area with relatively low charges for water service.[21] This result points to a need for more work on the role of prices in the treatment effect framework.

The significant coefficients in the expenditure equations with and without price knowledge are also similar in magnitude and sign. The only substantial difference is between the parameters on the household size variable. Both coefficients are positive, but the influence of larger households on water expenditures is greater when the price is not known. The water demand model results in Table 15 suggest that this result occurs because household size has a relatively greater effect on water demand without price knowledge. Therefore, unlike the result for lawn size, the variation in the influence of household size can be attributed to differences in behavior rather than differences in (exogenous) control characteristics.

The estimated variances of the spending outcome equations are significant, indicating the importance of unobserved heterogeneity in the sample. However, because of relatively insignificant correlations, the covariances between the price knowledge and spending equations are not significant. This implies that there is a limited amount of self-selection in this sample based on water price knowledge. Following the characterization in Chapter 2 and 3, the lack of significant covariance between price knowledge and spending decisions suggests that neither group has an 'absolute advantage' in their selected option. That is, price knowledge does not offer those who chose to know the price a significantly lower cost for utility than not knowing the price. A similar observation can be made for those who chose not to know the price.

---

[21] A casual comparison of mean lawn size and price variables by utility service area confirms that the two areas with the largest average lawn size also have the lowest mean charges.

The weaker condition of 'comparative advantage' applies when the average household with (without) price knowledge spends less for the same utility level than the average household without (with) price knowledge when they both (don't) know the price. A null hypothesis of no comparative advantage can be evaluated with a joint test of $\beta^{11} = \beta^{01}$ and $\sigma^{11D} = \sigma^{01D}$. The Wald statistic of 136.13 for these restrictions rejects joint equality with greater than 99% confidence. Thus, there is still significant information in the price knowledge (self-selection) decisions of households in the sample that can be used evaluate the relative valuations of price information. The treatment effect welfare measures introduced in Chapter 2 are designed to exploit this information.

The unconditional treatment effects and welfare measures for the value of price information are listed in Table 18. These figures are obtained by evaluating expressions (4-26) through (4-29) for each individual in the relevant group and averaging as defined in (4-30). The components of the policy relevant measure of compensating variation of price knowledge for the whole sample are shown in the first pane. A randomly selected household will spend $2.38 more per month if they know the price. However, this randomly chosen individual is not willing to pay a positive amount for the price knowledge. In fact, households are willing to pay –$1.42 on average for the price information so the extra spending for the price knowledge is not justified. This explains why over 95 percent of households reported not knowing the marginal price of service. The dominance of the households without price knowledge is evident in the last pane of Table 18. At $2.67 and -$1.47 per month, respectively, the estimated spending difference and willingness to pay for (perfect) price information for a randomly chosen household from this group are very close to the estimates for the sample shown in the first pane of Table 18.

Table 18. Bill treatment effects and welfare estimates of price information

Parameter	Bill Difference TEA
ATE	2.38
	(2.82)
E[CV]	-1.42
	(2.85)
TT	-1.78
	(1.96)
E[CV | know]	-0.69
	(1.97)
UT	2.67
	(2.87)
E[CV | don't know]	-1.47
	(2.89)

Standard deviations shown below the estimates.

The story is somewhat different for those who reported knowing the price. From the second pane in Table 18, this group actually spends less by $1.78 per month with price information, although they have a negative $0.69 willingness to pay for this information. However, households in this group still choose to know the price because it saves them money.

All results satisfy the treatment effects model consistency checks proposed in Chapter 2. The expected compensating variation is greater (less negative) than the spending difference (TT) for a randomly chosen household from the group that knows the price. This is not true for a randomly chosen household from those who do not know the price (UT) and from the sample in general (ATE). While the statistical significance of these measures was not tested directly, the rejection of joint equality in the comparative advantage test is suggestive. Note also that, although not shown, all of the consistency checks are met at the minimum and maximum values of the sample.

Discussion

This chapter investigated the differences in consumption behavior between households with and without information about the marginal price of public utility services. The analysis sought, in part, to determine whether or not measured *as-if* behavioral responses to price changes differ significantly from informed responses. A theoretical model of price perception was developed to formally review the comparative statics of price knowledge and the value of price information. The implications of the theoretical model were investigated with a structural water demand model and a treatment effect expenditures model based on the framework introduced in Chapter 2.

The estimation results from the simultaneous equations demand model with endogenous sample selection suggest that there are differences in average consumption, price elasticity, and perceived price between these two groups. Specifically, those who reported knowing the marginal price consumed less than average and were relatively more responsive to price variation. Those without price information behaved as if they were responding to a signal much lower than the actual marginal price. Consequently, this group was over-consuming public utility services relative to other goods and could possibly benefit by readjusting their budget to spend less on these services.

The opportunity cost of the over-expenditure on utility service constitutes the value of price information for those who do not know the price. This opportunity cost was estimated using the results from demand and expenditures models. The welfare measures of the value of price information in the demand model improve on existing approaches in considering the possibility that price knowledge changes the response to price changes.

The estimated average value of price information with the demand model is small relative to the average household's monthly income. This supports Shin's (1985) hypothesis that households do not know the price because the relative value of this information is relatively low. The results from treatment effects model suggest that the value of the information may be negative, but that some households would still learn the price because it saves them money. The upshot is that if public utilities (in the study area) make an effort to reduce the cost of obtaining accurate price information, then average consumption will decline as households who learn the marginal price adjust their budget allocations accordingly. This could be a point of interest for public utility managers interested in encouraging water conservation because it suggests that a relatively simple effort, such as clearly posting charges and usage information on water bills, could help achieve more efficient[22] levels of conservation from their residential customers.

---

[22] If water consumers know the actual marginal price for water service, then they can use this information to make efficient water conservation decisions with regard to utility maximization. However, the aggregate level of conservation achieved by the public utility will only be socially efficient to the degree that the marginal price for service reflects the full opportunity costs of providing that service (Carter and Milon 1999).

CHAPTER 5
SUMMARY

This dissertation explores ways to learn about values for public goods by observing the actions of individuals and households. Following an overview in the first chapter, Chapter 2 reviews conventional revealed preference valuation approaches and introduces the alternative *treatment effects approach* (TEA). Both the conventional and TEA approaches work by observing how expenditure patterns change with changes in the public good(s) of interest. However, these approaches differ in the way they operate when expenditure data is unavailable for the state of the world before or after the public good change.

Conventional methods estimate structural demand and/or utility equations for commodities or activities that are related to a public good of interest. With assumptions about the relationship between public and market goods, missing demand or utility outcomes are simulated as 'equivalent' price changes. Thus, prices are paramount in the conventional approaches to measuring public good values. Troubling aspects of conventional approaches' reliance on *activity-based* price indices are also reviewed in Chapter 2.

The TEA is an attempt to avoid some of the problems related to activity based price indices. It also uses expenditures for activities related to a public good. However, expenditures are not split into activity price and quantity indices to estimate a structural demand or utility model. This is because the TEA does not simulate missing expenditure outcomes related to a public good change using price changes. Rather, methods from the program evaluation literature are used to recover missing information using the observed expenditures of different groups of individuals. Those in a sample who use a public good for a specific activity are considered to be the *treatment group* and those who do not are considered to be the *control group*. Two ways are suggested for using such a *natural experiment* to derive values for public good access.

Chapter 3 applies a travel cost model and the TEA to measure the value of access to petroleum rigs in the Gulf of Mexico for recreational fishing. Both approaches are adapted to incorporate expenditures on fishing capital into welfare measures. The adaptation of the travel cost model turns out to be difficult because capital is not easily allocated at the trip level. However, the TEA readily incorporates outlays on capital equipment into the annual expenditure differences used to develop welfare measures.

The welfare measures calculationed for the travel cost model are unusually high because of a particularly low estimated coefficient on the trip cost variable. This suggests that other variations of the model should be estimated to examine the sensitivity of welfare measure estimates. Future work with this dataset should also include estimation of a conventional random utility model to generate alternative welfare measures of rig access.

The results for the TEA model estimation are promising. Annual measures of the value of rig access are 'reasonable' and the consistency tests proposed in Chapter 2 are satisfied. Specifically, the results indicated that rigs are not used by some individuals because they are not willing to pay the extra cost of rig use. This is true even though nonusers are willing to pay more on average than those who actually fish rigs. The inclusion of capital expenditures more than doubles the value of the TEA measures suggesting that these outlays are an important part of the opportunity cost of restricting rig fishing.

Chapter 4 uses a structural demand and a TEA models to examine price perception in the demand for public utility services and the value of price information. An analytical model is also developed to analyze price misperception and identify the value of perfect price information. Comparative statics results suggest that the price responsiveness is a function of price knowledge. Therefore, all else being equal, those who know the price will have a different price elasticity of demand than those who do not know the price.

The implications of the analytical model are analyzed with a water demand model and a TEA model of water expenditures estimated on a sample of households from north-central Florida. The sample is split into those who reported knowing the price and those who reported otherwise. The demand model results suggests that those who said they knew price had a larger price elasticity of demand than those who did not know the price. Furthermore, those who said they did not know the price were behaving as if water were free. The estimated value of price information for this group is low. The value is only slightly higher if price responsiveness is allowed to change with price knowledge.

Those who knew the price compose the treatment group and others form the control group for the TEA model. The differences in water bills between these two group are used to identify the expected value of price information. For most households in the sample, this value is low relative to the expected difference in spending associated with price knowledge. However, the results indicate that the group that reported knowing the price actually saved money by doing so.

APPENDIX
MATHEMATICA© DERRIVATION OF THE NET UTILITY FUNCTION

Following Hausman (1981), expenditure is characterized as a minimum value function that traces the minimum expenditure necessary to achieve a constant utility level given changes an exogenous variable. Hausman considers changes in a price variable, but I consider changes in the level or quality of an exogenously supplied public good z. Note that the public good indicator does not appear directly in the functional form chosen for the Engel equation:

```
px = a + g m[z] + c s + e
```

```
a + e + c s + g m[z]
```

where a is a constant, m is expenditure with parameter g, s is a vector of control variables with a conformable vector of parameters c, and e is an error term. Following Phlips (1983 p. 104), the underlying demand equation for this model can be defined as $x_i = a_j + b_i\, p_j/p_i + g\, m/p_i + e_i$. In this case the constant and error terms of the Engel equation are implicitly defined as $a = a_i\, p_i + b_i + p_j$ and $e = e_i\, p_i$,, respectively. This formulation treats prices as endogenous.

Following the approach in Hausman (1981), write the (uncompensated) Engel equation as a differential equation and solve for the expenditure as a function of the quantity or quality level of the public good:

```
mz = DSolve[m'[z] == px, m[z], z]
```

$$\left\{\left\{m[z] \to \frac{-a-e-c\,s}{g} + e^{g\,z}\, C[1]\right\}\right\}$$

Note that the derivation integrates with respect to the public good supply rather than prices since the price variables are assumed endogenous (unobservable) across the sample.

Solve for the constant of integration as the level of (indirect) utility:

[redacted]

Invert the indirect utility function in terms of income to get the expenditure function:

[redacted]

Use Roy's Identity and the indirect utility function to check the results:

[redacted]

Use Shephard's Lemma and the expenditure function to recover the compensated demand for the public good:

[redacted]

To derive the (indirect) utility difference to be used in the public good use selection equation first parameterize indirect utility functions for alternatives i and j:

[redacted]

$$-\frac{e^{-gi\,z}\,(-1+gi)\,(ai+ei+gi\,m+ci\,s)}{gi}$$

$$-\frac{e^{-gj\,z}\,(-1+gj)\,(aj+ej+gj\,m+cj\,s)}{gj}$$

Next, Calculate the difference in indirect utility from choosing alternative i over j:

```
ud = idui - iduj
```

$$-\frac{e^{-gi\,z}\,(-1+gi)\,(ai+ei+gi\,m+ci\,s)}{gi}+\frac{e^{-gj\,z}\,(-1+gj)\,(aj+ej+gj\,m+cj\,s)}{gj}$$

Collect variable terms:

```
udc = Collect[ud, {s, m}]
```

$$-\frac{ai\,e^{-gi\,z}\,(-1+gi)}{gi}-\frac{e^{-gi\,z}\,ei\,(-1+gi)}{gi}+\frac{aj\,e^{-gj\,z}\,(-1+gj)}{gj}+\frac{e^{-gj\,z}\,ej\,(-1+gj)}{gj}+\left(-e^{-gi\,z}\,(-1+gi)+e^{-gj\,z}\,(-1+gj)\right)m+\left(-\frac{ci\,e^{-gi\,z}\,(-1+gi)}{gi}+\frac{cj\,e^{-gj\,z}\,(-1+gj)}{gj}\right)s$$

Define a list of reduced form parameters for the net indirect utility function:

```
rep = {(-Exp[-gi z] (-1 + gi) +
     Exp[-gj z] (-1 + gj)) → G,
   (-ci Exp[-gi z] (-1 + gi) / gi +
     cj Exp[-gj z] (-1 + gj) / gj) → C,
   (-ai Exp[-gi z] (-1 + gi) / gi +
     aj Exp[-gj z] (-1 + gj) / gj) → A,
   (-ei Exp[-gi z] (-1 + gi) / gi +
     ej Exp[-gj z] (-1 + gj) / gj) → ed}
```

[illegible]

Finally, express the net indirect utility function in terms of the reduced form parameters:

redForm [illegible]

[illegible]

Note that the integration results in an argument for the public good condition appearing in as a multiple on the parameters of the reduced form net utility function. Consequently, I cannot constrain the parameters to be consistent across the index and spending equations as is done, for example, in the discrete/continuous models that combine hypothetical and observed data (Cameron 1992). It may be possible, however, to solve for the value of the public good indicator using the estimated reduced form parameters. This is not attempted in the present research.

REFERENCES

Agthe, D. E., R. B. Billings, J. L. Dobra, and K. Raffiee. "A Simultaneous Equation Demand Model for Block Rates." *Water Resources Research* 22, no. 1(1986): 1-4.

BEBR. "Survey of Northeast Florida Water Users." Bureau of Economic and Business Research. Gainesville: University of Florida, 1997.

Blundell, R., and M. Costa Dias. "Evaluation Methods for Non-Experimental Data." *Fiscal Studies* 21, no. 4(2000): 427-68.

Blundell, R., and T. Macurdy. "Labor Supply: A Review of Alternative Approaches." *Handbook of Labor Economics*. ed. O. Ashenfelter and D. Card, pp. 1559-1695. New York: Elsevier Science North-Holland, 1999.

Blundell, R., and J.-M. Robin. "Latent Separability: Grouping Goods without Weak Separability." *Econometrica* 68, no. 1(2000): 53-84.

Bockstael, N. E., and C. L. Kling. "Valuing Environmental Quality: Weak Complementarity with Sets of Goods." *American Journal of Agricultural Economics* 70, no. 3(1988): 654-662.

Bockstael, N. E., and K. E. McConnell. "Theory and Estimation of the Household Production Function for Wildlife Recreation." *Journal of Environmental Economics and Management* 8, no. 3(1981): 199-214.

---. "Welfare Measurement in the Household Production Framework." *The American Economic Review* 73, no. 4(1983): 806-814.

---. "Public Goods as Characteristics of Non-Market Commodities." *Economic Journal* 103, no. 420(1993): 1244-1257.

---. "The Behavioral Basis of Non-Market Valuation." *Valuing Recreation and the Environment*. ed. J. A. Herriges and C. L. Kling, pp. 1-32. Northampton: Elgar, 1999.

Bockstael, N. E., I. E. Strand, K. E. McConnell, and F. Arsanjani. "Sample Selection Bias in the Estimation of Recreation Demand Functions: An Application to Sportfishing." *Land Economics* 66, no. 1(1990): 40-49.

Bradford, D. F., and G. G. Hildebrandt. "Observable Preferences for Public Goods." *Journal of Public Economics* 8, no. 2(1977): 111-131.

Cameron, T. A. "Combining Contingent Valuation and Travel Cost Data for the Valuation of Nonmarket Goods." *Land Economics* 68, no. 3(1992): 302-17.

Carter, D. W., and J. W. Milon. "Marginal Opportunity Cost Vs. Average Cost Pricing of Water Service: Timing Issues for Pricing Reform." Staff Paper SP 98-19. Gainesville: Food and Resource Economics Department, University of Florida, 1998.

Carter, D. W., L. Perruso, and D. Lee. "Full Cost Accounting in Environmental Decision-Making." Extension Paper FE 310. Gainesville: Institute of Food and Agricultural Sciences, University of Florida, 2001.

Cavanagh, S. M., W. M. Hanemann, and R. N. Stavins. "Muffled Price Signals: Household Water Demand under Increasing-Block Prices." Working Paper. Cambridge: Environmental Economics Program at Harvard University, 2001.

Chicoine, D. L., S. C. Deller, and G. Ramamurthy. "Water Demand Estimation under Block Rate Pricing: A Simultaneous Equation Approach." *Water Resources Research* 22, no. 6(1986): 859-63.

Chicoine, D. L., and G. Ramamurthy. "Evidence on the Specification of Price in the Study of Domestic Water Demand." *Land Economics* 62, no. 1(1986): 28-32.

Conrad, K., and M. Schroder. "Demand for Durable and Nondurable Goods, Environmental Policy and Consumer Welfare." *Journal of Applied Econometrics* 6, no. 3(1991): 271-86.

Crooker, J., and C. L. Kling. "Nonparametric Bounds on Welfare Measures: A New Tool for Nonmarket Valuation." *Journal of Environmental Economics and Management* 39, no. 2(2000): 145-61.

Dauterive, L. "Rigs-to-Reefs Policy, Progress, and Perspective." OCS Report MMS 2000-073. New Orleans: U.S. Department of the Interior, Minerals Management Service, Gulf of Mexico Region, 2000.

Deaton, A., and J. Muellbauer. *Economics and Consumer Behavior*. New York: Cambridge University Press, 1980.

Deller, S. C., D. L. Chicoine, and G. Ramamurthy. "Instrumental Variables Approach to Rural Water Service Demand." *Southern Economic Journal* 53, no. 2(1986): 333-46.

Dobbs, I. M. "Adjusting for Sample Selection Bias in the Individual Travel Cost Method." *Journal of Agricultural Economics* 44, no. 2(1993): 335-342.

Dolton, P. J., and G. H. Makepeace. "Interpreting Sample Selection Effects." *Economics Letters* 24, no. 4(1987): 373-379.

Ebert, U. "Evaluation of Nonmarket Goods: Recovering Unconditional Preferences." *American Journal of Agricultural Economics* 80, no. 2(1998): 241-254.

Emerson, R. D. "Migratory Labor and Agriculture." *American Journal of Agricultural Economics* 71, no. 3(1989): 617-29.

Englin, J., and J. S. Shonkwiler. "Modeling Recreation Demand in the Presence of Unobservable Travel Costs: Toward a Travel Price Model." *Journal of Environmental Economics and Management* 29, no. 3(1995): 368-377.

English, D. B. K., and J. M. Bowker. "Sensitivity of Whitewater Rafting Consumers Surplus to Pecuniary Travel Cost Specifications." *Journal of Environmental Management* 47, no. 1(1996): 79-91.

Fix, P., J. Loomis, and R. Eichhorn. "Endogenously Chosen Travel Costs and the Travel Cost Model: An Application to Mountain Biking at Moab, Utah." *Applied Economics* 32, no. 10(2000): 1227-1231.

Frech, H. E., III. "Advertising as a Privately Supplied Public Good." *Economic Inquiry* 17, no. 3(1979): 414-418.

Freeman, A. M. *The Measurement of Environmental and Resource Values: Theory and Methods*. Washington, D.C.: Resources for the Future, 1993.

Frondel, M., and C. M. Schmidt. "Evaluating Environmental Programs: The Perspective of Modern Evaluation Research." Discussion Paper 397. Bohn: Institute for the Study of Labor, 2001.

Gourieroux, C. *Econometrics of Qualitative Dependent Variables*. New York: Cambridge University Press, 2000.

Greene, W. H. *Limdep Version 7.0 User's Manual*. Bellport: Econometric Software, Inc., 1995.

---. *Econometric Analysis*. Upper Saddle River: Prentice Hall, 2000.

Griffin, R. C., and C. Chang. "Pretest Analyses of Water Demand in Thirty Communities." *Water Resources Research* 26, no. 10(1990): 2251-2255.

Haab, T. C., and R. L. Hicks. "Accounting for Choice Set Endogeneity in Random Utility Models of Recreation Demand." *Journal of Environmental Economics and Management* 34, no. 2(1997): 127-147.

Haab, T. C., and K. E. McConnell. "Count Data Models and the Problem of Zeros in Recreation Demand Analysis." *American Journal of Agricultural Economics* 78, no. 1(1996): 89-102.

---. *Valuing Environmental and Natural Resources: The Econometrics of Non-Market Valuation*. Northampton: Edward Elgar, 2002.

Hanemann, M. "Determinants of Urban Water Use." *Urban Water Demand Management and Planning*. ed. D. D. Baumann and J. J. Boland. New York: McGraw-Hill, 1998.

Hanemann, M., and E. Morey. "Separability, Partial Demand Systems, and Consumer's Surplus Measures." *Journal of Environmental Economics and Management* 22, no. 3(1992): 241-258.

Hanemann, W. M. "Measuring the Worth of Natural Resource Facilities: Comment." *Land Economics* 56, no. 4(1980): 482-86.

---. "Discrete-Continuous Models of Consumer Demand." *Econometrica* 52, no. 3(1984a): 541-561.

---. "Welfare Evaluations in Contingent Valuation Experiments with Discrete Responses." *American Journal of Agricultural Economics* 66, no. 3(1984b): 332-41.

---. "Welfare Analysis with Discrete Choice Models." *Valuing Recreation and the Environment: Revealed Preference Methods in Theory and Practice*. ed. J. A. Herriges and C. L. Kling, pp. 33-64. Northampton: Edward Elgar, 1999.

Hau, T. D.-K. "A Hicksian Approach to Cost-Benefit Analysis with Discrete-Choice Models." *Economica* 52, no. 28(1985): 479-90.

Hauber, A. B., and G. R. Parsons. "The Effect of Nesting Structure Specification on Welfare Estimation in a Random Utility Model of Recreation Demand: An Application to the Demand for Recreational Fishing." *American Journal of Agricultural Economics* 82, no. 3(2000): 501-14.

Hausman, J. A. "Exact Consumer's Surplus and Deadweight Loss." *American Economic Review* 71, no. 4(1981): 662-676.

---. "The Econometrics of Nonlinear Budget Sets." *Econometrica* 53, no. 6(1985): 1255-82.

Heckman, J. J. "Dummy Endogenous Variables in a Simultaneous Equation System." *Econometrica* 46, no. 4(1978): 931-59.

---. "Instrumental Variables: A Study of Implicit Behavioral Assumptions Used in Making Program Evaluations." *Journal of Human Resources* 32, no. 3(1997): 441-462.

---. "Accounting for Heterogeneity, Diversity and General Equilibrium in Evaluating Social Programmes." *Economic Journal* 111, no. 475(2001a): F654-99.

---. "Micro Data, Heterogeneity, and the Evaluation of Public Policy: Nobel Lecture." *Journal of Political Economy* 109, no. 4(2001b): 673-748.

Heckman, J. J., N. Hohmann, and J. Smith. "Substitution and Dropout Bias in Social Experiments: A Study of an Influential Social Experiment." *Quarterly Journal of Economics* 115, no. 2(2000): 651-694.

Heckman, J. J., and R. Robb, Jr. "Alternative Methods for Evaluating the Impact of Interventions: An Overview." *Journal of Econometrics* 30, no. 1-2(1985): 239-67.

Heckman, J. J., J. L. Tobias, and E. J. Vytlacil. "Four Parameters of Interest in the Evaluation of Social Programs." *Southern Economic Journal* 68, no. 2(2001): 210-223.

Heckman, J. J., and E. J. Vytlacil. "The Relationship between Treatment Parameters within a Latent Variable Framework." *Economics Letters* 66, no. 1(2000): 33-39.

---. "Local Instrumental Variables." *Nonlinear Statistical Modeling*. ed. C. Hsiao, K. Morimune, and J. L. Powell, pp. 1-48. Cambridge: Cambridge University Press, 2001a.

---. "Policy-Relevant Treatment Effects." *American Economic Review* 91, no. 2(2001b): 107-111.

Hellerstein, D. "Can We Count on Count Models?" *Valuing Recreation and the Environment: Revealed Preference Methods in Theory and Practice*. ed. J. A. Herriges and C. L. Kling, pp. 271-83. Northampton: Edward Elgar, 1999.

Herriges, J. A., and K. K. King. "Residential Demand for Electricity under Inverted Block Rates: Evidence from a Controlled Experiment." *Journal of Business and Economic Statistics* 12, no. 4(1994): 419-30.

Herriges, J. A., and C. L. Kling. *Valuing Recreation and the Environment: Revealed Preference Methods in Theory and Practice*. Northampton: Edward Elgar, 1999.

Herriges, J. A., C. L. Kling, and D. J. Phaneuf. "Corner Solution Models of Recreation Demand: A Comparison of Competing Frameworks." *Valuing Recreation and the Environment: Revealed Preference Methods in Theory and Practice*. ed. J. A. Herriges and C. L. Kling, pp. 163-97. Northampton: Edward Elgar, 1999.

Hewitt, J. A. "A Discrete/Continuous Choice Approach to Residential Water Demand under Block Rate Pricing: Reply." *Land Economics* 76, no. 2(2000): 324-30.

Hewitt, J. A., and W. M. Hanemann. "A Discrete/Continuous Choice Approach to Residential Water Demand under Block Rate Pricing." *Land Economics* 71, no. 2(1995): 173-92.

Houthakker, H. S., P. K. Verleger, Jr., and D. P. Sheehan. "Dynamic Demand Analyses for Gasoline and Residential Electricity." *American Journal of Agricultural Economics* 56, no. 2(1974): 412-18.

Johnansson, P.-O. *The Economic Theory and Measurement of Environmental Benefits*. New York: Cambridge University Press, 1991.

---. *Cost-Benefit Analysis of Environmental Change*. New York: Cambridge University Press, 1993.

Jones, C. V., and J. R. Morris. "Instrumental Price Estimates and Residential Water Demand." *Water Resources Research* 20, no. 2(1984): 197-202.

Kling, C. L. "A Note on the Welfare Effects of Omitting Substitute Prices and Qualities from Travel Cost Models." *Land Economics* 65, no. 3(1989): 290-96.

Kling, C. L., and C. J. Thomson. "The Implications of Model Specification for Welfare Estimation in Nested Logit Models." *American Journal of Agricultural Economics* 78, no. 1(1996): 103-114.

Kolodinsky, J. "Time as a Direct Source of Utility: The Case of Price Information Search for Groceries." *Journal of Consumer Affairs* 24, no. 1(1990): 89-109.

LaFrance, J. T., and W. M. Hanemann. "The Dual Structure of Incomplete Demand Systems." *American Journal of Agricultural Economics* 71, no. 2(1989): 262-74.

Laitila, T. "Estimation of Combined Site-Choice and Trip-Frequency Models of Recreational Demand Using Choice-Based and on-Site Samples." *Economics Letters* 64, no. 1(1999): 17-23.

Larson, D. M., and S. L. Shaikh. "Empirical Specification Requirements for Two-Constraint Models of Recreation Choice." *American Journal of Agricultural Economics* 83, no. 2(2001): 428-40.

Lee, L.-f., G. S. Maddala, and R. P. Trost. "Asymptotic Covariance Matrices of Two-Stage Probit and Two-Stage Tobit Methods for Simultaneous Equations Models with Selectivity." *Econometrica* 48, no. 2(1980): 491-503.

Lee, L.-F., and M. M. Pitt. "Microeconometric Demand Systems with Binding Nonnegativity Constraints: The Dual Approach." *Econometrica* 54, no. 5(1986): 1237-42.

Loehman, E. "Alternative Measures of Benefit for Nonmarket Goods Which Are Substitutes or Complements for Market Goods." *Social Choice and Welfare* 8, no. 4(1991): 275-305.

Maddala, G. S. *Limited-Dependent and Qualitative Variables in Econometrics*. Cambridge: Cambridge University Press, 1983.

Maddock, R., E. Castano, and F. Vella. "Estimating Electricity Demand: The Cost of Linearising the Budget Constraint." *Review of Economics and Statistics* 74, no. 2(1992): 350-54.

Maler, K.-G. *Environmental Economics: A Theoretical Inquiry*. Baltimore: The John Hopkins University Press, 1974.

Moffitt, R. "The Econometrics of Piecewise-Linear Budget Constraints: A Survey and Exposition of the Maximum Likelihood Method." *Journal of Business and Economic Statistics* 4, no. 3(1986): 317-28.

---. "The Econometrics of Kinked Budget Constraints." *Journal of Economic Perspectives* 4, no. 2(1990): 119-39.

Moffitt, R. A. "Models of Treatment Effects When Responses Are Heterogeneous. Commentary." *Proceedings of the National Academy of Sciences of the United States of America* 96, no. 12(1998): 6575-6576.

Morey, E. R., W. S. Breffle, and P. A. Greene. "Two Nested Constant-Elasticity-of-Substitution Models of Recreational Participation and Site Choice: An "Alternatives" Model and an "Expenditures" Model." *American Journal of Agricultural Economics* 83, no. 2(2001): 414-427.

Nieswiadomy, M. L. "Estimating Urban Residential Water Demand: Effects of Price Structure, Conservation, and Education." *Water Resources Research* 28, no. 3(1992): 609-15.

Nieswiadomy, M. L., and D. J. Molina. "Comparing Residential Water Demand Estimates under Decreasing and Increasing Block Rates Using Household Data." *Land Economics* 65, no. 3(1989): 280-89.

---. "A Note of Price Perception in Water Demand Models." *Land Economics* 67, no. 3(1991): 352-59.

Parsons, G. R., and A. B. Hauber. "Spatial Boundaries and Choice Set Definition in a Random Utility Model of Recreation Demand." *Land Economics* 74, no. 1(1998): 32-48.

Parsons, G. R., P. M. Jakus, and T. Tomasi. "A Comparison of Welfare Estimates from Four Models for Linking Seasonal Recreational Trips to Multinomial Logit Models of Site Choice." *Journal of Environmental Economics and Management* 38, no. 2(1999): 143-157.

Parsons, G. R., and M. J. Kealy. "Randomly Drawn Opportunity Sets in a Random Utility Model of Lake Recreation." *Land Economics* 68, no. 1(1992): 93-106.

Parsons, G. R., and M. S. Needelman. "Site Aggregation in a Random Utility Model of Recreation." *Land Economics* 68, no. 4(1992): 418-33.

Parsons, G. R., A. J. Plantinga, and K. J. Boyle. "Narrow Choice Sets in a Random Utility Model of Recreation Demand." *Land Economics* 76, no. 1(2000): 86-99.

Phaneuf, D. J. "A Dual Approach to Modeling Corner Solutions in Recreation Demand." *Journal of Environmental Economics and Management* 37, no. 1(1999): 85-105.

Phaneuf, D. J., C. L. Kling, and J. A. Herriges. "Valuing Water Quality Improvements Using Revealed Preference Methods When Corner Solutions Are Present." *American Journal of Agricultural Economics* 80, no. 5(1998): 1025-1031.

---. "Estimation and Welfare Calculations in a Generalized Corner Solution Model with an Application to Recreation Demand." *Review of Economics and Statistics* 82, no. 1(2000): 83-92.

Phlips, L. *Applied Consumption Analysis*. New York: Amsterdam North-Holland Pub. Co., 1983.

Pollak, R. A. "Conditional Demand Functions and Consumption Theory." *Quarterly Journal of Economics* 83, no. 1(1969): 60-78.

---. "Price Dependent Preferences." *American Economic Review* 67, no. 2(1977): 64-75.

Pollak, R. A., and M. L. Wachter. "The Relevance of the Household Production Function and Its Implications for the Allocation of Time." *Journal of Political Economy* 83, no. 2(1975): 255-77.

Quantech. "Recreational Uses of Oil and Gas Structures in the Gulf of Mexico." Report to U.S. Mineral Management Service. Arlington, VA, 2001.

Randall, A. *Resource Economics: An Economic Approach to Natural Resource and Environmental Policy*. New York: Wiley, 1987.

---. "A Difficulty with the Travel Cost Method." *Land Economics* 70, no. 1(1994): 88-96.

Reiss, P. C. "Household Electricity Demand, Revisited." Working Paper 8687. Cambridge: National Bureau of Economic Research, 2001.

Rietveld, P., J. Rouwendal, and B. Zwart. "Estimating Water Demand in Indonesia: A Maximum Likelihood Approach to Block Rate Pricing Data." Tinbergen Institute Discussion Paper TI 97-072/3. Amsterdam: Free University, 1997.

Rosenthal, D. H. "The Necessity for Substitute Prices in Recreation Demand Analyses." *American Journal of Agricultural Economics* 69, no. 4(1987): 828-837.

Shapiro, P., and T. Smith. "Preferences for Nonmarket Goods Revealed through Market Demands." *Advances in Applied Microeconomics*. ed. V. K. Smith, pp. 105-122: JAI Press Inc., 1981.

Shaw, D. "On-Site Samples' Regression: Problems of Non-Negative Integers, Truncation, and Endogenous Stratification." *Journal of Econometrics* 37, no. 2(1988): 211-223.

Shaw, W. D., and P. Feather. "Possibilities for Including the Opportunity Cost of Time in Recreation Demand Systems." *Land Economics* 75, no. 4(1999): 592-602.

Shaw, W. D., and J. S. Shonkwiler. "Brand Choice and Purchase Frequency Revisited: An Application to Recreation Behavior." *American Journal of Agricultural Economics* 82, no. 3(2000): 515-526.

Shechter, M. "A Comparative Study of Environmental Amenity Valuations." *Environmental and Resource Economics* 1, no. 2(1991): 129-55.

Shin, J.-S. "Perception of Price When Price Information Is Costly: Evidence from Residential Electricity Demand." *Review of Economics and Statistics* 67, no. 4(1985): 591-98.

Smith, V. K. "The Influence of Resource and Environmental Problems on Applied Welfare Economics: An Introductory Essay." *Environmental Resources and Applied Welfare Economics*. ed. V. K. Smith, pp. 3-43. Washington D.C.: Resources for the Future, 1988a.

---. "Selection and Recreation Demand." *American Journal of Agricultural Economics* 70, no. 1(1988b): 29-36.

---. "Welfare Effects, Omitted Variables, and the Extent of the Market." *Land Economics* 69, no. 2(1993): 121-31.

---. *Estimating Economic Values for Nature: Methods for Non-Market Valuation*. Brookfield: Edward Elgar, 1996.

Terza, J. V. "Determinants of Household Electricity Demand: A Two-Stage Probit Approach." *Southern Economic Journal* 52, no. 4(1986): 1131-39.

Terza, J. V., and W. P. Welch. "Estimating Demand under Block Rates: Electricity and Water." *Land Economics* 58, no. 2(1982): 181-88.

Varian, H. R. *Microeconomic Analysis, 3rd Edition*. New York: W.W. Norton & Co., 1992.

Vella, F., and M. Verbeek. "Estimating and Interpreting Models with Endogenous Treatment Effects." *Journal of Business and Economic Statistics* 17, no. 4(1999): 473-478.

Wales, T. J., and A. D. Woodland. "Estimation of Consumer Demand Systems with Binding Non-Negativity Constraints." *Journal of Econometrics* 21, no. 3(1983): 263-85.

Ward, F. A. "Specification Considerations for the Price Variable in Travel Cost Demand Models." *Land Economics* 60, no. 3(1984): 301-05.

Whitehead, J. C., T. C. Haab, and J.-C. Huang. "Measuring Recreation Benefits of Quality Improvements with Revealed and Stated Behavior Data." *Resource and Energy Economics* 22, no. 4(2000): 339-54.

Wilman, E. A., and R. J. Pauls. "Sensitivity of Consumers' Surplus Estimates to Variation in the Parameters of the Travel Cost Model." *Canadian Journal of Agricultural Economics* 35, no. 1(1987): 197-212.

Winship, C., and S. L. Morgan. "The Estimation of Causal Effects from Observational Data." *Annual Review of Sociology* 25, no. 1(1999): 659-706.

Zerbe, R. O., and D. D. Dively. *Benefit-Cost Analysis in Theory and Practice*. New York: Harper Collins College Pubishers, 1994.

Ziemer, R. F., W. N. Musser, F. C. White, and C. Hill. "Sample Selection Bias in Analysis of Consumer Choice: An Application to Warmwater Fishing Demand Outdoor Recreation." *Water Resources Research* 18, no. 2(1982): 215-221.

BIOGRAPHICAL SKETCH

David Carter was born in Columbus, Ohio USA and promptly moved to Florida to grow up; A task he is still working on. David attended schools in the Tampa Bay area before enrolling in Stetson University in Deland, Florida. After receiving a degree in economics he began his nearly eight year career as a Gator at the University of Florida. This career as a professional student culminated in a M.S. and, now, a Ph.D. from the Food and Resource Economics Department. David plans to continue studying choices related to the natural environment in Florida and beyond.

I certify that I have read this study and that in my opinion it conforms to acceptable standards of scholarly presentation and is fully adequate, in scope and quality, as a dissertation for the degree of Doctor of Philosophy.

J. Walter Milon, Chair
Professor of Food and Resource Economics

I certify that I have read this study and that in my opinion it conforms to acceptable standards of scholarly presentation and is fully adequate, in scope and quality, as a dissertation for the degree of Doctor of Philosophy.

Clyde F. Kiker, Cochair
Professor of Food and Resource Economics

I certify that I have read this study and that in my opinion it conforms to acceptable standards of scholarly presentation and is fully adequate, in scope and quality, as a dissertation for the degree of Doctor of Philosophy.

Robert D. Emerson
Professor of Food and Resource Economics

I certify that I have read this study and that in my opinion it conforms to acceptable standards of scholarly presentation and is fully adequate, in scope and quality, as a dissertation for the degree of Doctor of Philosophy.

Donna J. Lee
Associate Professor of Food and Resource Economics

I certify that I have read this study and that in my opinion it conforms to acceptable standards of scholarly presentation and is fully adequate, in scope and quality, as a dissertation for the degree of Doctor of Philosophy.

Lawrence W. Kenny
Professor of Economics

This dissertation was submitted to the Graduate Faculty of the College of Agricultural and Life Sciences and to the Graduate School and was accepted as partial fulfillment of the requirements for the degree of Doctor of Philosophy.

December 2002

Dean. College of Agricultural and Life Sciences

Dean. Graduate School

Printed by Libri Plureos GmbH in Hamburg,
Germany

9 780530 000671